Organometallic Chemistry

Organometallic Chemistry

Edited by
Eugene Pressley

Larsen & Keller
www.larsen-keller.com

Organometallic Chemistry
Edited by Eugene Pressley
ISBN: 978-1-63549-675-8 (Hardback)

© 2018 Larsen & Keller

 Larsen & Keller

Published by Larsen and Keller Education,
5 Penn Plaza,
19th Floor,
New York, NY 10001, USA

Cataloging-in-Publication Data

Organometallic chemistry / edited by Eugene Pressley.
      p. cm.
Includes bibliographical references and index.
ISBN 978-1-63549-675-8
      1. Organometallic chemistry. 2. Organometallic compounds. 3. Chemistry, Organic.
I. Pressley, Eugene.
QD411 .O74 2018
547.05--dc23

For more information regarding Larsen and Keller Education and its products, please visit the publisher's website www.larsen-keller.com

Table of Contents

Preface

Organometallic chemistry refers to that branch of chemistry that studies the chemical compounds, which have one bond between carbon atom and a metal. It includes the study of alkaline Earth, alkaline and transition metal. This subject uses the elements of both organic and inorganic chemistry. Organometallic compounds are primarily used as homogeneous catalysis agents in industries. The topics covered in this book offer the readers new insights in the field of organometallic chemistry. Also included in it is a detailed explanation of the various concepts and applications of the field. For all those who are interested in this subject, this textbook can prove to be an essential guide.

A short introduction to every chapter is written below to provide an overview of the content of the book:

Chapter 1 - Organometallic chemistry is a branch of inorganic chemistry that studies compounds formed between carbon and a metal. The characterization of these compounds can be ascertained after a complete evaluation of the compounds on the basis of structure-property paradigm. This chapter is an overview of the subject matter incorporating all the major aspects of organometallic chemistry; **Chapter 2** - Elements in the periodic table are categorized into s, p, d and f blocks depending on the type of atomic orbital that have. Most organometallic compounds can be prepared by methods of synthesis of metals reacting with organic halide or through the process of hydrometallation, metathesis, and metal displacement. The major aspects of the elements in organometallic chemistry are discussed in this section; **Chapter 3** - The p-Block metals can lose electrons easily and form organometallic compounds. Their reaction pattern includes oxidation, Lewis acidity, and its nucleophilic nature. The chapter closely examines the key concepts of organometallic chemistry to provide an extensive understanding of the subject; **Chapter 4** - The topics discussed in the chapter are of great importance to broaden the existing knowledge on the compound in organometallic chemistry. The main factors affecting the reactions of organometallic compounds along with reaction patterns have been discussed in the following section. It has been carefully written to provide an easy understanding of the subject matter; **Chapter 5** - Metal alkyls are readily available on Earth and are a major source of stabilized carbanions. These alkyls can be divided into two categories – stable alkyls and agostic alkyls. The section also explores hyrides and its facets. Hydrides have assumed prominence in organometallic chemistry due to its ability to undergo insertion reaction and form various organometallic compounds. The nature of hydrogen in hydrides can vary from being protic to hydidic. This section has been carefully written to provide an easy understanding of the varied facets of alkyls and hydrides; **Chapter 6** - Carbonyl is a fundamental group with double-bonded carbon atom with oxygen. The elements containing this group are referred to as carbonyl compounds. Phosphine is a colorless, odourless, flammable and toxic gas with the chemical formula NH_3. The topics elaborated in this chapter will help in gaining a better perspective about carbonyls, phosphines and the concept of substitution; **Chapter 7** - Through oxidative addition, two anionic ligands and added to the metal centre and its coordination number and oxidation state is increased. Reductive elimination is the opposite of oxidative addition, through which two ligands are removed from the metal centre. It forms a covalent bond between ligands and leads to the decrease in the oxidation state the metal centre. The section also deals with insertion and elimination reaction, and nucleophilic and electrophilic addition and abstraction. All the diverse principles of organometallic chemistry have been carefully analyzed in this chapter; **Chapter 8 -** Catalysis is the increase in the rate of chemical reaction caused due to the introduction of a substance called catalyst. The substance, however, does not directly participate in the process, but helps speed up the reaction. Catalysis can be classified as homogenous and heterogeneous. The former plays a vital role in application of organometallic chemistry. It is the form of catalysis in which the catalyst is of the same phase as the reactants. The topics elaborated in this chapter will help in gaining a better perspective about catalysis in organometallic chemistry.

I extend my sincere thanks to the publisher for considering me worthy of this task. Finally, I thank my family for being a source of support and help.

Editor

Fundamentals of Organometallic Chemistry

Organometallic chemistry is a branch of inorganic chemistry that studies compounds formed between carbon and a metal. The characterization of these compounds can be ascertained after a complete evaluation of the compounds on the basis of structure-property paradigm. This chapter is an overview of the subject matter incorporating all the major aspects of organometallic chemistry.

Organometallic Chemistry

Organometallic chemistry is the study of chemical compounds containing at least one bond between a carbon atom of an organic compound and a metal, including alkaline, alkaline earth, transition metal, and other cases. Moreover, some related compounds such as transition metal hydrides and metal phosphine complexes are often included in discussions of organometallic compounds. The field of organometallic chemistry combines aspects of traditional inorganic and organic chemistry.

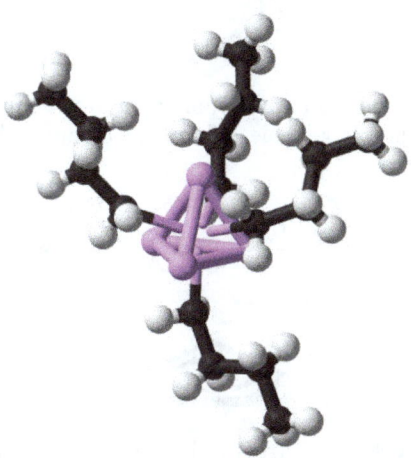

n-Butyllithium, an organometallic compound. Four lithium atoms (in purple) form a tetrahedron, with four butyl groups attached to the faces (carbon is black, hydrogen is white).

Organometallic compounds are widely used both stoichiometrically in research and industrial chemical reactions, as well as in the role of catalysts to increase the rates of such reactions (e.g., as in uses of homogeneous catalysis), where target molecules include polymers, pharmaceuticals, and many other types of practical products.

Organometallic Compounds

Organometallic compounds are distinguished by the prefix "organo-" e.g. organopalladium compounds. Examples of such organometallic compounds include all Gilman reagents, which contain lithium and copper. Tetracarbonyl nickel, and ferrocene are examples of organometallic compounds containing transition metals. Other examples include organomagnesium compounds like iodo(methyl)magnesium MeMgI, dimethylmagnesium (Me_2Mg), and all Grignard reagents; organolithium compounds such as n-butyllithium (n-BuLi), organozinc compounds such as diethylzinc (Et_2Zn) and chloro(ethoxycarbonylmethyl)zinc ($ClZnCH_2C(=O)OEt$); and organocopper compounds such as lithium dimethylcuprate ($Li^+[CuMe_2]^-$).

The term "metalorganics" usually refers to metal-containing compounds lacking direct metal-carbon bonds but which contain organic ligands. Metal beta-diketonates, alkoxides, and dialkylamides are representative members of this class.

In addition to the traditional metals, lanthanides, actinides, and semimetals, elements such as boron, silicon, arsenic, and selenium are considered to form organometallic compounds, e.g. organoborane compounds such as triethylborane (Et_3B).

- Representative Organometallic Compounds

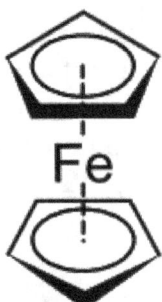

Ferrocene is an archetypal organoiron complex.
It is an air-stable, sublimable compound.

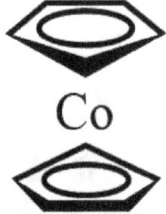

Cobaltocene is a structural analogue of ferrocene,
but is highly reactive toward air.

Tris(triphenylphosphine)rhodium carbonyl hydride is used in the commercial
production of many aldehyde-based fragrances.

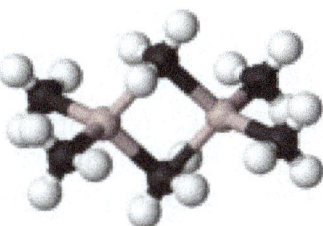

Zeise's salt is an example of a transition metal alkene complex.

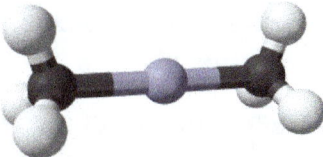

Trimethylaluminium is an organometallic compound with a bridgingmethyl group.
It is used in the industrial production of some alcohols

Dimethylzinc has a linear coordination. It is a volatile pyrophoric liquid
that is used in the preparation of semiconducting films.

Coordination Compounds with Organic Ligands

Many complexes feature coordination bonds between a metal and organic ligands. The organic ligands often bind the metal through a heteroatom such as oxygen or nitrogen, in which case such compounds are considered coordination compounds. However, if any of the ligands form a direct M-C bond, then complex is usually considered to be organometallic, e.g., $[(C_6H_6)Ru(H_2O)_3]^{2+}$. Furthermore, many lipophilic compounds such as metal acetylacetonates and metal alkoxides are called "metalorganics."

Many organic coordination compounds occur naturally. For example, hemoglobin and myoglobin contain an iron center coordinated to the nitrogen atoms of a porphyrin ring; magnesium is the center of a chlorin ring in chlorophyll. The field of such inorganic compounds is known as bioinorganic chemistry. In contrast to these coordination compounds, methylcobalamin (a form of Vitamin B_{12}), with a cobalt-methyl bond, is a true organometallic complex, one of the few known in biology. This subset of complexes are often discussed within the subfield of bioorganometallic chemistry. Illustrative of the many functions of the B_{12}-dependent enzymes, the MTR enzyme catalyzes the transfer of a methyl group from a nitrogen on N5-methyl-tetrahydrofolate to the sulfur of homocysteine to produce methionine.

The status of compounds in which the canonical anion has a delocalized structure in which the negative charge is shared with an atom more electronegative than carbon, as in enolates, may vary with the nature of the anionic moiety, the metal ion, and possibly the medium; in the absence of direct structural evidence for a carbon–metal bond, such compounds are not considered to be organometallic.

Structure and Properties

The metal-carbon bond in organometallic compounds are generally highly covalent. For highly electropositive elements, such as lithium and sodium, the carbon ligand exhibits carbanionic character, but free carbon-based anions are extremely rare, an example being cyanide.

Concepts and Techniques

As in other areas of chemistry, electron counting is useful for organizing organometallic chemistry. The 18-electron rule is helpful in predicting the stabilities of metal carbonyls and related compounds. Most organometallic compounds do not however follow the 18e rule. Chemical bonding and reactivity in organometallic compounds is often discussed from the perspective of the isolobal principle.

As well as X-ray diffraction, NMR and infrared spectroscopy are common techniques used to determine structure. The dynamic properties of organometallic compounds is often probed with variable-temperature NMR and chemical kinetics.

Organometallic compounds undergo several important reactions:

- oxidative addition and reductive elimination
- transmetalation
- carbometalation
- hydrometalation
- electron transfer
- beta-hydride elimination
- organometallic substitution reaction
- carbon-hydrogen bond activation
- cyclometalation
- migratory insertion
- nucleophilic abstraction

History

Early developments in organometallic chemistry include Louis Claude Cadet's synthesis of methyl arsenic compounds related to cacodyl, William Christopher Zeise'splatinum-ethylene complex, Edward Frankland's discovery of dimethyl zinc, Ludwig Mond's discovery of $Ni(CO)_4$, and Victor

Grignard's organomagnesium compounds. The abundant and diverse products from coal and petroleum led to Ziegler-Natta, Fischer-Tropsch, hydroformylation catalysis which employ CO, H_2, and alkenes as feedstocks and ligands.

Recognition of organometallic chemistry as a distinct subfield culminated in the Nobel Prizes to Ernst Fischer and Geoffrey Wilkinson for work on metallocenes. In 2005, Yves Chauvin, Robert H. Grubbs and Richard R. Schrock shared the Nobel Prize for metal-catalyzed olefin metathesis.

Organometallic Chemistry Timeline

- 1760 Louis Claude Cadet de Gassicourt investigates inks based on cobalt salts and isolates cacodyl from cobalt mineral containing arsenic.

- 1827 William Christopher Zeise produces Zeise's salt; the first platinum / olefin complex.

- 1848 Edward Frankland discovers diethylzinc.

- 1863 Charles Friedel and James Crafts prepare organochlorosilanes.

- 1890 Ludwig Mond discovers nickel carbonyl.

- 1899 Introduction of Grignard reaction.

- 1899 John Ulric Nef discovers alkynation using sodium acetylides.

- 1900 Paul Sabatier works on hydrogenation organic compounds with metal catalysts. Hydrogenation of fats kicks off advances in food industry.

- 1909 Paul Ehrlich introduces Salvarsan for the treatment of syphilis, an early arsenic based organometallic compound.

- 1912 Nobel PrizeVictor Grignard and Paul Sabatier.

- 1930 Henry Gilman works on lithium cuprates.

- 1951 Walter Hieber was awarded the Alfred Stock prize for his work with metal carbonyl chemistry.

- 1951 Ferrocene is discovered.

- 1963 Nobel prize for Karl Ziegler and Giulio Natta on Ziegler-Natta catalyst.

- 1965 Discovery of cyclobutadieneiron tricarbonyl.

- 1968 Heck reaction.

- 1973 Nobel prizeGeoffrey Wilkinson and Ernst Otto Fischer on sandwich compounds.

- 1981 Nobel prizeRoald Hoffmann and Kenichi Fukui for creation of the Woodward-Hoffman Rules.

- 2001 Nobel prizeW. S. Knowles, R. Noyori and Karl Barry Sharpless for asymmetric hydrogenation.

- 2005 Nobel prizeYves Chauvin, Robert Grubbs, and Richard Schrock on metal-catalyzed alkene metathesis.

- 2010 Nobel prizeRichard F. Heck, Ei-ichi Negishi, Akira Suzuki for palladium catalyzed cross coupling reactions.

Scope

Subspecialty areas of organometallic chemistry include:

- Period 2 elements: organolithium chemistry, organoberyllium chemistry, organoborane chemistry.

- Period 3 elements: organomagnesium chemistry, organoaluminum chemistry, organosilicon chemistry.

- Period 4 elements: organotitanium chemistry, organochromium chemistry, organomanganese chemistryorganoiron chemistry, organocobalt chemistryorganonickel chemistry, organocopper chemistry, organozinc chemistry, organogallium chemistry, organogermanium chemistry.

- Period 5 elements: organoruthenium chemistry, organopalladium chemistry, organosilver chemistry, organocadmium chemistry, organoindium chemistry, organotin chemistry.

- Period 6 elements: organolanthanide chemistry, organoosmium chemistry, organoiridium chemistry, organoplatinum chemistry, organogold chemistry, organomercury chemistry, organothallium chemistry, organolead chemistry.

- Period 7 elements: organouranium chemistry.

The following is a presentation of elements of the periodic table with known compounds of carbon with other elements.

CH																	He
CLi	CBe											CB	CC	CN	CO	CF	Ne
CNa	CMg											CAl	CSi	CP	CS	CCl	CAr
CK	CCa	CSc	CTi	CV	CCr	CMn	CFe	CCo	CNi	CCu	CZn	CGa	CGe	CAs	CSe	CBr	CKr
CRb	CSr	CY	CZr	CNb	CMo	CTc	CRu	CRh	CPd	CAg	CCd	CIn	CSn	CSb	CTe	CI	CXe
CCs	CBa		CHf	CTa	CW	CRe	COs	CIr	CPt	CAu	CHg	CTl	CPb	CBi	CPo	CAt	Rn
Fr	CRa		Rf	Db	CSg	Bh	Hs	Mt	Ds	Rg	Cn	Nh	Fl	Mc	Lv	Ts	Og

CLa	CCe	CPr	CNd	CPm	CSm	CEu	CGd	CTb	CDy	CHo	CEr	CTm	CYb	CLu
Ac	CTh	CPa	CU	CNp	CPu	CAm	CCm	CBk	CCf	CEs	Fm	Md	No	Lr

Chemical bonds to carbon	
Core organic chemistry	Many uses in chemistry
Academic research, but no widespread use	Bond unknown

Industrial Applications

Organometallic compounds find wide use in commercial reactions, both as homogeneous catalysis and as stoichiometric reagents For instance, organolithium, organomagnesium, and organoaluminium compounds, examples of which are highly basic and highly reducing, are useful stoichiometrically, but also catalyze many polymerization reactions.

Almost all processes involving carbon monoxide rely on catalysts, notable examples being described as carbonylations. The production of acetic acid from methanol and carbon monoxide is catalyzed via metal carbonyl complexes in the Monsanto process and Cativa process. Most synthetic aldehydes are produced via hydroformylation. The bulk of the synthetic alcohols, at least those larger than ethanol, are produced by hydrogenation of hydroformylation-derived aldehydes. Similarly, the Wacker process is used in the oxidation of ethylene to acetaldehyde.

Almost all industrial processes involving alkene-derived polymers rely on organometallic catalysts. The world's polyethylene and polypropylene are produced via both heterogeneously via Ziegler-Natta catalysis and homogeneously, e.g., via constrained geometry catalysts.

Most processes involving hydrogen rely on metal-based catalysts. Whereas bulk hydrogenations, e.g. margarine production, rely on heterogeneous catalysts, For the production of fine chemicals, such hydrogenations rely on soluble organometallic complexes or involve organometallic intermediates. Organometallic complexes allow these hydrogenations to be effected asymmetrically.

A constrained geometry organotitanium complex is a precatalyst for olefin polymerization.

Many semiconductors are produced from trimethylgallium, trimethylindium, trimethylaluminium, and trimethylantimony. These volatile compounds are decomposed along with ammonia, arsine, phosphine and related hydrides on a heated substrate via metalorganic vapor phase epitaxy (MOVPE) process in the production of light-emitting diodes (LEDs).

Environmental Concerns

Natural and contaminant organometallic compounds are found in the environment. Some that are remnants of human use, such as organolead and organomercury compounds, are toxicity hazards. Tetraethyllead was prepared for use as a gasoline additive but has fallen into disuse because of lead's toxicity. Its replacements are other organometallic compounds, such as ferrocene and methylcyclopentadienyl manganese tricarbonyl (MMT). The organoarsenic compound roxarsone is a controversial animal feed additive. In 2006, approximately one million kilograms of it were produced in the U.S alone.

Roxarsone is an organoarsenic compound used as an animal feed.

Characterization of Organometallic Complexes

The characterization of an organometallic complex involves obtaining a complete understanding of the same right from its identification to the assessment of its purity content, to even elucidation of its stereochemical features. Detailed structural understanding of the organometallic compounds is critical for obtaining an insight on its properties and which is achieved based on the structure-property paradigm.

Synthesis and Isolation

Synthesis and isolation are two very important experimental protocols in the overall scheme of things of organometallic chemistry and thus these needs to be performed carefully. The isolation of the organometallic compounds is essential for their characterization and reactivity studies. Fortunately, many of the methods of organic chemistry can be used in organometallic chemistry as the organometallic compounds are mostly nonvolatile crystalline solids at room temperature and atmospheric pressure though a few examples of these compounds are known to exist in the liquid $[(CH_3C_5H_4Mn(CO)_3]$ and even in the vapor $[Ni(CO)_4]$ states. The organometallic compounds are comparatively more sensitive to aerial oxygen and moisture, and because of which the manipulation of these compounds requires stringent experimental skills to constantly provide them with anaerobic environment for their protection. All of these necessities led to the development of the so-called special Schlenk techniques, requiring special glasswares and which in conjunction with a high vacuum line and a dry box allow the lab bench-top manipulation of these compounds. Successful isolation of organometallic compounds naturally points to the need for various spectroscopic techniques for their characterizations and some of the important ones are discussed below.

^{1}H NMR Spectroscopy

The ^{1}H NMR spectroscopy is among the extensively used techniques for the characterization of organometallic compounds. Of particular interest is the application of ^{1}H NMR spectroscopy in the characterization of the metal hydride complexes, for which the metal hydride moiety appear at a distinct chemical shift range between 0 ppm to −40 ppm to the high field of tetramethyl silane (TMS). This upfield shift of the metal hydride moiety is attributed to a shielding by metal d−electrons and the extent of the upfield shift increases with higher the d^n configuration. Chemical shifts, peak intensities as well as coupling constants from the through-bond couplings between adjacent nuclei like that of the observation of J_{P-H}, if a phosphorous nucleus is present within the coupling range of a proton nucleus, are often used for the analysis of these compounds. The ^{1}H NMR spectroscopy is often successfully employed in studying more complex issues like fluxionality and diastereotopy in organometallic molecules.

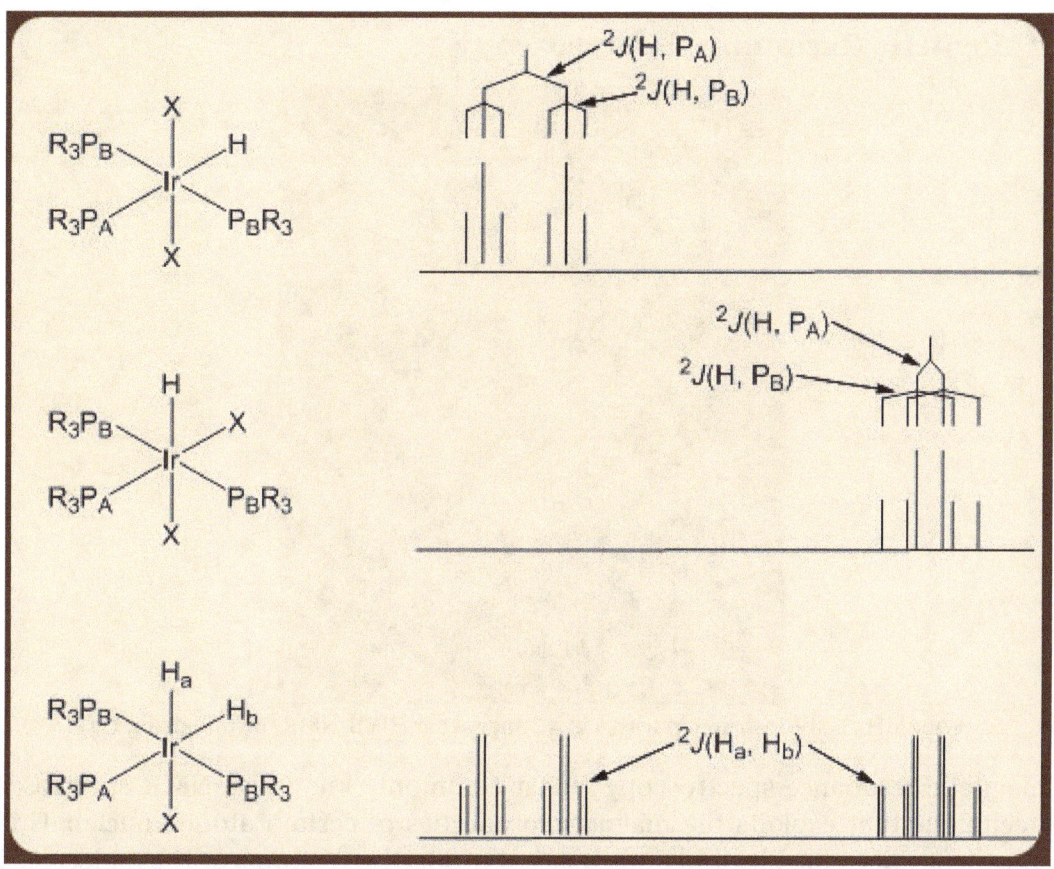

Different phosphorous-proton coupling patterns in various iridium hydride complexes.

The paramagnetic organometallic complexes show a large range of chemical shifts, for example, $(\eta^6-C_6H_6)_2V$ exhibits proton resonances that extend even up to 290 ppm.

13C NMR Spectroscopy

Although the natural abundance of NMR active ^{13}C (I = ½) nuclei is only 1 %, it is possible to obtain a proton decoupled $^{13}C\{^1H\}$ NMR spectra for most of the organometallic complexes. In addition, the off–resonance ^1H decoupled ^{13}C experiments yield $^1J_{C-H}$ coupling constants, which contain vital structural information, and hence are very critical to the ^{13}C NMR spectral analysis. For example, the $^1J_{C-H}$ coupling constants directly correlate with the hybridization of the C–H bonds with sp center exhibiting a $^1J_{C-H}$ coupling constant of ~250 Hz, a sp^2 center of 160 Hz and a sp^3 center of 125 Hz. Similar to what is seen in ^1H NMR, a phosphorous–carbon coupling is also observed in a ^{13}C NMR spectrum with the *trans* coupling (~100 Hz) being larger than the *cis* coupling (~10 Hz).

31P NMR Spectroscopy

The ^{31}P NMR spectroscopy, which in conjunction with ^1H and ^{13}C NMR spectroscopies, is a useful technique in studying the phosphine containing organometallic complexes. The ^{31}P NMR experiments are routinely run under ^1H decoupled conditions for simplification of the spectral features that allow convenience in spectral analysis. Thus, for this very reason, many mechanistic studies on catalytic cycle are conveniently undertaken by ^{31}P NMR spectroscopy whenever applicable.

Nuclear Magnetic Resonance Spectroscopy

A 900MHz NMR instrument with a 21.1 T magnet at HWB-NMR, Birmingham, UK

Nuclear magnetic resonance spectroscopy', most commonly known as NMR spectroscopy, is a research technique that exploits the magnetic properties of certain atomic nuclei. This type of spectroscopy determines the physical and chemical properties of atoms or the molecules in which they are contained. It relies on the phenomenon of nuclear magnetic resonance and can provide detailed information about the structure, dynamics, reaction state, and chemical environment of molecules. The intramolecular magnetic field around an atom in a molecule changes the resonance frequency, thus giving access to details of the electronic structure of a molecule and its individual functional groups.

Most frequently, NMR spectroscopy is used by chemists and biochemists to investigate the properties of organic molecules, although it is applicable to any kind of sample that contains nuclei possessing spin. Suitable samples range from small compounds analyzed with 1-dimensional proton or carbon-13 NMR spectroscopy to large proteins or nucleic acids using 3 or 4-dimensional techniques. The impact of NMR spectroscopy on the sciences has been substantial because of the range of information and the diversity of samples, including solutions and solids.

NMR spectra are unique, well-resolved, analytically tractable and often highly predictable for small molecules. Thus, in organic chemistry practice, NMR analysis is used to confirm the identity of a substance. Different functional groups are obviously distinguishable, and identical functional groups with differing neighboring substituents still give distinguishable signals. NMR has largely replaced traditional wet chemistry tests such as color reagents or typical chromatography for identification. A disadvantage is that a relatively large amount, 2–50 mg, of a purified substance is required, although it may be recovered through a workup. Preferably, the sample should be dissolved in a solvent, because NMR analysis of solids requires a dedicated MAS machine and may not give equally well-resolved spectra. The timescale of NMR is relatively long, and thus it is not suitable for observing fast phenomena, producing only an averaged spectrum. Although large amounts of impurities do show on an NMR spectrum, better methods exist for detecting impurities, as NMR is inherently not very sensitive - though at higher frequencies, sensitivity narrows.

NMR spectrometers are relatively expensive; universities usually have them, but they are less common in private companies. Modern NMR spectrometers have a very strong, large and expensive liquid helium-cooled superconducting magnet, because resolution directly depends on magnetic field strength. Less expensive machines using permanent magnets and lower resolution are also available, which still give sufficient performance for certain application such as reaction monitoring and quick checking of samples. There are even benchtop NMR spectrometers.

History

The Purcell group at Harvard University and the Bloch group at Stanford University independently developed NMR spectroscopy in the late 1940s and early 1950s. Edward Mills Purcell and Felix Bloch shared the 1952 Nobel Prize in Physics for their discoveries.

Basic NMR Techniques

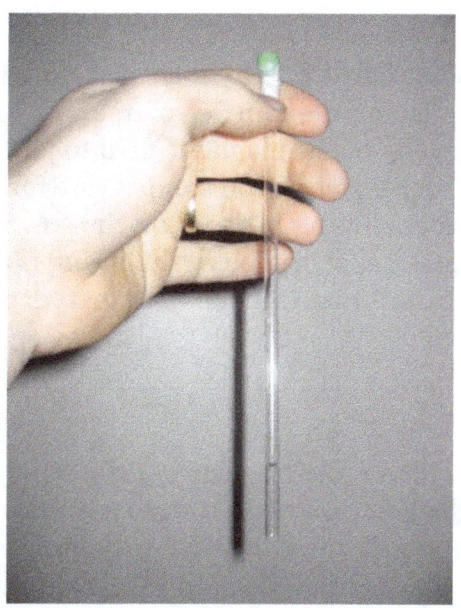

The NMR sample is prepared in a thin-walled glass tube - an NMR tube.

Resonant Frequency

When placed in a magnetic field, NMR active nuclei (such as ^1H or ^{13}C) absorb electromagnetic radiation at a frequency characteristic of the isotope. The resonant frequency, energy of the absorption, and the intensity of the signal are proportional to the strength of the magnetic field. For example, in a 21 Tesla magnetic field, protons resonate at 900 MHz. It is common to refer to a 21 T magnet as a 900 MHz magnet, although different nuclei resonate at a different frequency at this field strength in proportion to their nuclear magnetic moments.

Sample Handling

A NMR spectrometer typically consists of a spinning sample-holder inside a very strong magnet, a radio-frequency emitter and a receiver with a probe (an antenna assembly) that goes inside the magnet to surround the sample, optionally gradient coils for diffusion measurements and electronics to control the system. Spinning the sample is necessary to average out diffusional motion. Whereas, measurements of diffusion constants (*diffusion ordered spectroscopy* or DOSY) are done the sample stationary and spinning off, and flow cells can be used for online analysis of process flows.

Deuterated Solvents

The vast majority of nuclei in a solution would belong to the solvent, and most regular solvents are hydrocarbons and would contain NMR-reactive protons. Thus, deuterium (hydrogen-2) is substituted (99+%). The most used deuterated solvent is deuterochloroform ($CDCl_3$), although deuterium oxide (D_2O) and deuterated DMSO (DMSO-d_6) are used for hydrophilic analytes. The chemical shifts are slightly different in different solvents, depending on electronic solvation effects. NMR spectra are often calibrated against the known solvent residual proton peak instead of added tetramethylsilane.

Shim and Lock

To detect the very small frequency shifts due to nuclear magnetic resonance, the applied magnetic field must be constant throughout the sample volume. High resolution NMR spectrometers use shims to adjust the homogeneity of the magnetic field to parts per billion (ppb) in a volume of a few cubic centimeters. In order to detect and compensate for inhomogeneity and drift in the magnetic field, the spectrometer maintains a "lock" on the solvent deuterium frequency with a separate lock unit. In modern NMR spectrometers shimming is adjusted automatically, though in some cases the operator has to optimize the shim parameters manually to obtain the best possible resolution.

Acquisition of Spectra

Upon excitation of the sample with a radio frequency (60–1000 MHz) pulse, a nuclear magnetic resonance response - a free induction decay (FID) - is obtained. It is a very weak signal, and requires sensitive radio receivers to pick up. A Fourier transform is carried out to extract the frequency-domain spectrum from the raw time-domain FID. A spectrum from a single FID has a low signal-to-noise ratio, but fortunately it improves readily with averaging of repeated acquisitions. Good ^1H NMR spectra can be acquired with 16 repeats, which takes only minutes. However, for elements heavier than hydrogen, the relaxation time is rather long, e.g. around 8 seconds for ^{13}C. Thus, acquisition of quantitative heavy-element spectra can be time-consuming, taking tens of minutes to hours.

If the second excitation pulse is sent prematurely before the relaxation is complete, the average magnetization vector still points in a nonparallel direction, giving suboptimal absorption and emission of the pulse. In practice, the peak areas are then not proportional to the stoichiometry; only the presence, but not the amount of functional groups is possible to discern. An inversion recovery experiment can be done to determine the relaxation time and thus the required delay between pulses. A 180° pulse, an adjustable delay, and a 90° pulse is transmitted. When the 90° pulse exactly cancels out the signal, the delay corresponds to the time needed for 90° of relaxation. Inversion recovery is worthwhile for quantitive ^{13}C, 2D and other time-consuming experiments.

Chemical Shift

A spinning charge generates a magnetic field that results in a magnetic moment proportional to the spin. In the presence of an external magnetic field, two spin states exist (for a spin 1/2 nucleus): one spin up and one spin down, where one aligns with the magnetic field and the other opposes it. The difference in energy (ΔE) between the two spin states increases as the strength of the field increases, but this difference is usually very small, leading to the requirement for strong NMR magnets (1-20 T for modern NMR instruments). Irradiation of the sample with energy corresponding to the exact spin state separation of a specific set of nuclei will cause excitation of those set of nuclei in the lower energy state to the higher energy state.

For spin 1/2 nuclei, the energy difference between the two spin states at a given magnetic field strength is proportional to their magnetic moment. However, even if all protons have the same magnetic moments, they do not give resonant signals at the same frequency values. This difference arises from the differing electronic environments of the nucleus of interest. Upon application of an external magnetic field, these electrons move in response to the field and generate local magnetic fields that oppose the much stronger applied field. This local field thus "shields" the proton from the applied magnetic field, which must therefore be increased in order to achieve resonance (absorption of rf energy). Such increments are very small, usually in parts per million (ppm). For instance, the proton peak from an aldehyde is shifted ca. 10 ppm compared to a hydrocarbon peak, since as an electron-withdrawing group, the carbonyl deshields the proton by reducing the local electron density. The difference between 2.3487 T and 2.3488 T is therefore about 42 ppm. However a frequency scale is commonly used to designate the NMR signals, even though the spectrometer may operate by sweeping the magnetic field, and thus the 42 ppm is 4200 Hz for a 100 MHz reference frequency (rf).

However given that the location of different NMR signals is dependent on the external magnetic field strength and the reference frequency, the signals are usually reported relative to a reference signal, usually that of TMS (tetramethylsilane). Additionally, since the distribution of NMR signals is field dependent, these frequencies are divided by the spectrometer frequency. However, since we are dividing Hz by MHz, the resulting number would be too small, and thus it is multiplied by a million. This operation therefore gives a locator number called the "chemical shift" with units of parts per million. In general, chemical shifts for protons are highly predictable since the shifts are primarily determined by simpler shielding effects (electron density), but the chemical shifts for many heavier nuclei are more strongly influenced by other factors including excited states ("paramagnetic" contribution to shielding tensor).

The chemical shift provides information about the structure of the molecule. The conversion of the raw data to this information is called *assigning* the spectrum. For example, for the ^{1}H-NMR spectrum for ethanol (CH_3CH_2OH), one would expect signals at each of three specific chemical shifts: one for the CH_3 group, one for the CH_2 group and one for the OH group. A typical CH$_3$ group has a shift around 1 ppm, a CH$_2$ attached to an OH has a shift of around 4 ppm and an OH has a shift anywhere from 2–6 ppm depending on the solvent used and the amount of hydrogen bonding. While the O atom does draw electron density away from the attached H through their mutual sigma bond, the electron lone pairs on the O bathe the H in their shielding effect.

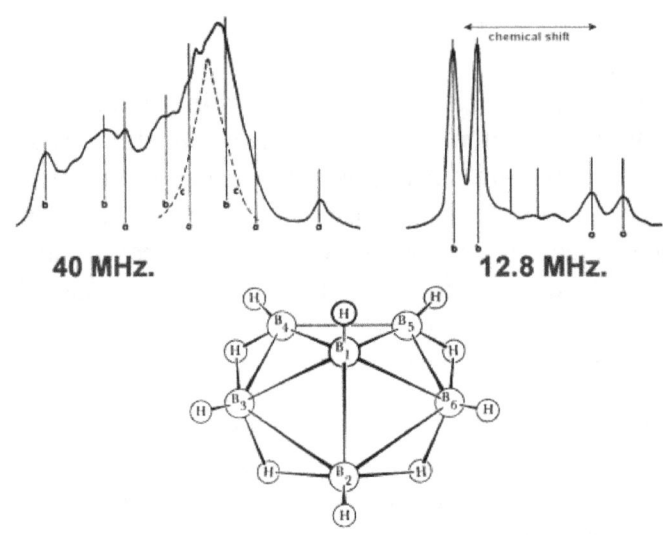

40 MHz.   **12.8 MHz.**

Example of the chemical shift: NMR spectrum of hexaborane B$_6$H$_{10}$ showing peaks shifted in frequency, which give clues as to the molecular structure. (click to read interpretation details)

In Paramagnetic NMR spectroscopy, measurements are conducted on paramagnetic samples. The paramagnetism gives rise to very diverse chemical shifts. In 1H NMR spectroscopy, the chemical shift range can span 500 ppm.

Because of molecular motion at room temperature, the three methyl protons *average out* during the NMR experiment (which typically requires a few ms). These protons become degenerate and form a peak at the same chemical shift.

The shape and area of peaks are indicators of chemical structure too. In the example above—the proton spectrum of ethanol—the CH$_3$ peak has three times the area as the OH peak. Similarly the CH$_2$ peak would be twice the area of the OH peak but only 2/3 the area of the CH$_3$ peak.

Software allows analysis of signal intensity of peaks, which under conditions of optimal relaxation, correlate with the number of protons of that type. Analysis of signal intensity is done by integration—the mathematical process that calculates the area under a curve. The analyst must integrate the peak and not measure its height because the peaks also have *width*—and thus its size is dependent on its area not its height. However, it should be mentioned that the number of protons, or any other observed nucleus, is only proportional to the intensity, or the integral, of the NMR signal in the very simplest one-dimensional NMR experiments. In more elaborate experiments, for instance, experiments typically used to obtain carbon-13 NMR spectra, the integral of the signals depends on the relaxation rate of the nucleus, and its scalar and dipolar coupling constants. Very often these factors are poorly known - therefore, the integral of the NMR signal is very difficult to interpret in more complicated NMR experiments.

J-coupling

Multiplicity	Intensity Ratio
Singlet (s)	1
Doublet (d)	1:1
Triplet (t)	1:2:1
Quartet (q)	1:3:3:1
Quintet	1:4:6:4:1
Sextet	1:5:10:10:5:1
Septet	1:6:15:20:15:6:1

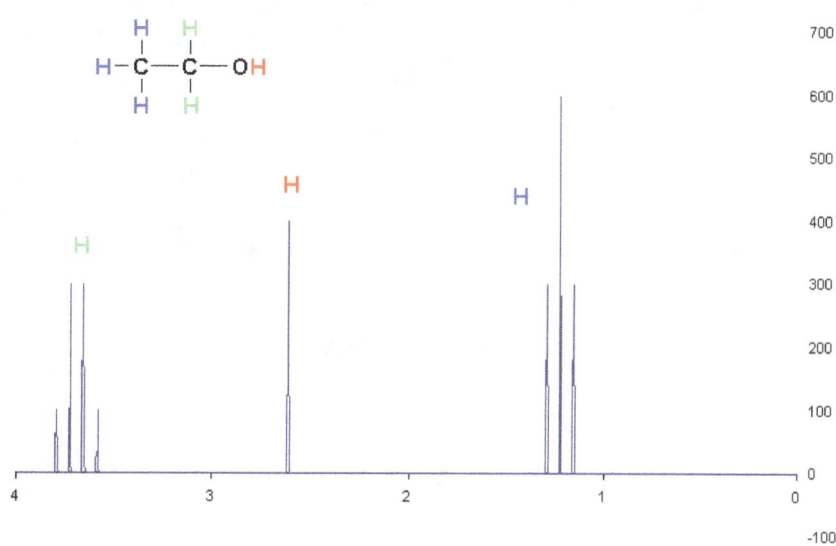

Example ^{1}H NMR spectrum (1-dimensional) of ethanol plotted as signal intensity vs. chemical shift. There are three different types of H atoms in ethanol regarding NMR. The hydrogen (H) on the -OH group is not coupling with the other H atoms and appears as a singlet, but the CH_3- and the -CH_2- hydrogens are coupling with each other, resulting in a triplet and quartet respectively.

Some of the most useful information for structure determination in a one-dimensional NMR spectrum comes from J-coupling or scalar coupling (a special case of spin-spin coupling) between NMR active nuclei. This coupling arises from the interaction of different spin states through the chemical bonds of a molecule and results in the splitting of NMR signals. These splitting patterns can be complex or simple and, likewise, can be straightforwardly interpretable or deceptive. This coupling provides detailed insight into the connectivity of atoms in a molecule.

Coupling to n equivalent (spin ½) nuclei splits the signal into a $n+1$ multiplet with intensity ratios following Pascal's triangle as described on the right. Coupling to additional spins will lead to further splittings of each component of the multiplet e.g. coupling to two different spin ½ nuclei with significantly different coupling constants will lead to a *doublet of doublets* (abbreviation: dd). Note that coupling between nuclei that are chemically equivalent (that is, have the same chemical shift) has no effect on the NMR spectra and couplings between nuclei that are distant (usually more than 3 bonds apart for protons in flexible molecules) are usually too small to cause observable splittings. *Long-range* couplings over more than three bonds can often be observed in cyclic and aromatic compounds, leading to more complex splitting patterns.

For example, in the proton spectrum for ethanol described above, the CH$_3$ group is split into a *triplet* with an intensity ratio of 1:2:1 by the two neighboring CH$_2$ protons. Similarly, the CH$_2$ is split into a *quartet* with an intensity ratio of 1:3:3:1 by the three neighboring CH$_3$ protons. In principle, the two CH$_2$ protons would also be split again into a *doublet* to form a *doublet of quartets* by the hydroxyl proton, but intermolecular exchange of the acidic hydroxyl proton often results in a loss of coupling information.

Coupling to any spin ½ nuclei such as phosphorus-31 or fluorine-19 works in this fashion (although the magnitudes of the coupling constants may be very different). But the splitting patterns differ from those described above for nuclei with spin greater than ½ because the spin quantum number has more than two possible values. For instance, coupling to deuterium (a spin 1 nucleus) splits the signal into a *1:1:1 triplet* because the spin 1 has three spin states. Similarly, a spin 3/2 nucleus splits a signal into a *1:1:1:1 quartet* and so on.

Coupling combined with the chemical shift (and the integration for protons) tells us not only about the chemical environment of the nuclei, but also the number of *neighboring* NMR active nuclei within the molecule. In more complex spectra with multiple peaks at similar chemical shifts or in spectra of nuclei other than hydrogen, coupling is often the only way to distinguish different nuclei.

1D PROTON SPECTRUM

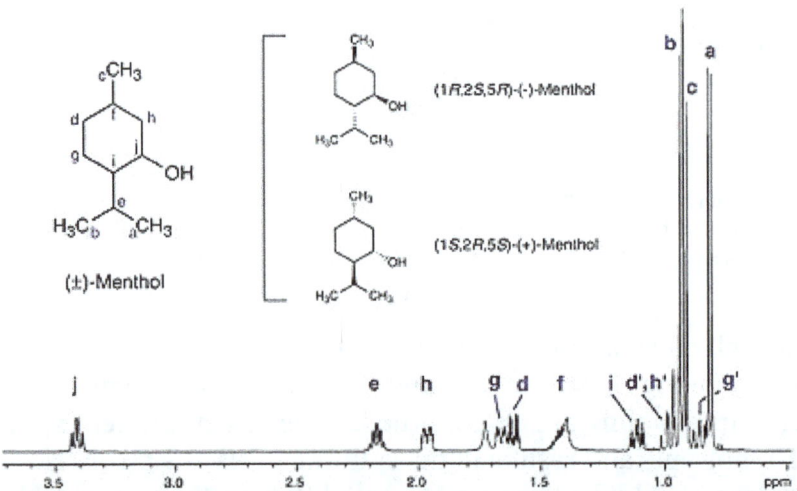

^{1}H NMR spectrum of menthol with chemical shift in ppm on the horizontal axis. Each magnetically inequivalent proton has a characteristic shift, and couplings to other protons appear as splitting of the peaks into multiplets: e.g. peak *a*, because of the three magnetically equivalent protons in methyl group *a*, couple to one adjacent proton (*e*) and thus appears as a doublet.

Second-order (or Strong) Coupling

The above description assumes that the coupling constant is small in comparison with the difference in NMR frequencies between the inequivalent spins. If the shift separation decreases (or the coupling strength increases), the multiplet intensity patterns are first distorted, and then become more complex and less easily analyzed (especially if more than two spins are involved). Intensification of some peaks in a multiplet is achieved at the expense of the remainder, which sometimes almost disappear in the background noise, although the integrated area under the peaks remains constant. In most high-field NMR, however, the distortions are usually modest and the characteristic distortions (*roofing*) can in fact help to identify related peaks.

Some of these patterns can be analyzed with the method published by John Pople, though it has limited scope.

Second-order effects decrease as the frequency difference between multiplets increases, so that high-field (i.e. high-frequency) NMR spectra display less distortion than lower frequency spectra. Early spectra at 60 MHz were more prone to distortion than spectra from later machines typically operating at frequencies at 200 MHz or above.

Magnetic Inequivalence

More subtle effects can occur if chemically equivalent spins (i.e., nuclei related by symmetry and so having the same NMR frequency) have different coupling relationships to external spins. Spins that are chemically equivalent but are not indistinguishable (based on their coupling relationships) are termed magnetically inequivalent. For example, the 4 H sites of 1,2-dichlorobenzene divide into two chemically equivalent pairs by symmetry, but an individual member of one of the pairs has different couplings to the spins making up the other pair. Magnetic inequivalence can lead to highly complex spectra which can only be analyzed by computational modeling. Such effects are more common in NMR spectra of aromatic and other non-flexible systems, while conformational averaging about C-C bonds in flexible molecules tends to equalize the couplings between protons on adjacent carbons, reducing problems with magnetic inequivalence.

Correlation Spectroscopy

Correlation spectroscopy is one of several types of two-dimensional nuclear magnetic resonance (NMR) spectroscopy or 2D-NMR. This type of NMR experiment is best known by its acronym, COSY. Other types of two-dimensional NMR include J-spectroscopy, exchange spectroscopy (EXSY), Nuclear Overhauser effect spectroscopy (NOESY), total correlation spectroscopy (TOCSY) and heteronuclear correlation experiments, such as HSQC, HMQC, and HMBC. In correlation spectroscopy, emission is centered on the peak of an individual nucleus; if its magnetic field is correlated with another nucleus by through-bond (COSY, HSQC, etc.) or through-space (NOE) coupling, a response can also be detected on the frequency of the correlated nucleus. Two-dimensional NMR spectra provide more information about a molecule than one-dimensional NMR spectra and are especially useful in determining the structure of a molecule, particularly for molecules that are too complicated to work with using one-dimensional NMR. The first two-dimensional experiment, COSY, was proposed by Jean Jeener, a professor at Université Libre de Bruxelles, in 1971. This experiment was later implemented by Walter P. Aue, Enrico Bartholdi and Richard R. Ernst, who published their work in 1976.

Solid-state Nuclear Magnetic Resonance

A variety of physical circumstances do not allow molecules to be studied in solution, and at the same time not by other spectroscopic techniques to an atomic level, either. In solid-phase media, such as crystals, microcrystalline powders, gels, anisotropic solutions, etc., it is in particular the dipolar coupling and chemical shift anisotropy that become dominant to the behaviour of the nuclear spin systems. In conventional solution-state NMR spectroscopy, these additional interactions would lead to a significant broadening of spectral lines. A variety of techniques allows establishing high-resolution conditions, that can, at least for ^{13}C spectra, be comparable to solution-state NMR spectra.

Two important concepts for high-resolution solid-state NMR spectroscopy are the limitation of possible molecular orientation by sample orientation, and the reduction of anisotropic nuclear magnetic interactions by sample spinning. Of the latter approach, fast spinning around the magic angle is a very prominent method, when the system comprises spin 1/2 nuclei. Spinning rates of ca. 20 kHz are used, which demands special equipment. A number of intermediate techniques, with samples of partial alignment or reduced mobility, is currently being used in NMR spectroscopy.

Applications in which solid-state NMR effects occur are often related to structure investigations on membrane proteins, protein fibrils or all kinds of polymers, and chemical analysis in inorganic chemistry, but also include "exotic" applications like the plant leaves and fuel cells. For example, Rahmani et al. studied the effect of pressure and temperature on the bicellar structures' self-assembly using deuterium NMR spectroscopy.

Biomolecular NMR Spectroscopy

Proteins

Much of the innovation within NMR spectroscopy has been within the field of proteinNMR spectroscopy, an important technique in structural biology. A common goal of these investigations is to obtain high resolution 3-dimensional structures of the protein, similar to what can be achieved by X-ray crystallography. In contrast to X-ray crystallography, NMR spectroscopy is usually limited to proteins smaller than 35 kDa, although larger structures have been solved. NMR spectroscopy is often the only way to obtain high resolution information on partially or wholly intrinsically unstructured proteins. It is now a common tool for the determination of Conformation Activity Relationships where the structure before and after interaction with, for example, a drug candidate is compared to its known biochemical activity. Proteins are orders of magnitude larger than the small organic molecules, but the basic NMR techniques and some NMR theory also applies. Because of the much higher number of atoms present in a protein molecule in comparison with a small organic compound, the basic 1D spectra become crowded with overlapping signals to an extent where direct spectral analysis becomes untenable. Therefore, multidimensional (2, 3 or 4D) experiments have been devised to deal with this problem. To facilitate these experiments, it is desirable to isotopically label the protein with ^{13}C and ^{15}N because the predominant naturally occurring isotope ^{12}C is not NMR-active and the nuclear quadrupole moment of the predominant naturally occurring ^{14}N isotope prevents high resolution information from being obtained from this nitrogen isotope. The most important method used for structure determination of proteins utilizes NOE experiments to measure distances between pairs of atoms within the molecule. Subsequently, the distances obtained are used to generate a 3D structure of the molecule by solving a distance geometry problem. NMR can also be used to obtain information on the dynamics and conformational flexibility of different regions of a protein.

Nucleic Acids

"Nucleic acid NMR" is the use of NMR spectroscopy to obtain information about the structure and dynamics of polynucleic acids, such as DNA or RNA. As of 2003, nearly half of all known RNA structures had been determined by NMR spectroscopy.

Nucleic acid and protein NMR spectroscopy are similar but differences exist. Nucleic acids have a smaller percentage of hydrogen atoms, which are the atoms usually observed in NMR spectroscopy, and because nucleic acid double helices are stiff and roughly linear, they do not fold back on themselves to give "long-range" correlations. The types of NMR usually done with nucleic acids are ^{1}H or proton NMR, ^{13}C NMR, ^{15}N NMR, and ^{31}P NMR. Two-dimensional NMR methods are almost always used, such as correlation spectroscopy (COSY) and total coherence transfer spectroscopy (TOCSY) to detect through-bond nuclear couplings, and nuclear Overhauser effect spectroscopy (NOESY) to detect couplings between nuclei that are close to each other in space.

Parameters taken from the spectrum, mainly NOESY cross-peaks and coupling constants, can be used to determine local structural features such as glycosidic bond angles, dihedral angles (using the Karplus equation), and sugar pucker conformations. For large-scale structure, these local parameters must be supplemented with other structural assumptions or models, because errors add up as the double helix is traversed, and unlike with proteins, the double helix does not have a compact interior and does not fold back upon itself. NMR is also useful for investigating nonstandard geometries such as bent helices, non-Watson–Crick basepairing, and coaxial stacking. It has been especially useful in probing the structure of natural RNA oligonucleotides, which tend to adopt complex conformations such as stem-loops and pseudoknots. NMR is also useful for probing the binding of nucleic acid molecules to other molecules, such as proteins or drugs, by seeing which resonances are shifted upon binding of the other molecule.

Carbohydrates

Carbohydrate NMR spectroscopy addresses questions on the structure and conformation of carbohydrates.

Infrared Spectroscopy

Infrared spectroscopy (IR spectroscopy or Vibrational Spectroscopy) involves the interaction of infrared radiation with matter. It covers a range of techniques, mostly based on absorption spectroscopy. As with all spectroscopic techniques, it can be used to identify and study chemicals. For a given sample which may be solid, liquid, or gaseous, the method or technique of infrared spectroscopy uses an instrument called an infrared spectrometer (or spectrophotometer) to produce an infrared spectrum. A basic IR spectrum is essentially a graph of infrared light absorbance (or transmittance) on the vertical axis vs. frequency or wavelength on the horizontal axis. Typical units of frequency used in IR spectra are reciprocal centimeters (sometimes called wave numbers), with the symbol cm^{-1}. Units of IR wavelength are commonly given in micrometers (formerly called "microns"), symbol µm, which are related to wave numbers in a reciprocal way. A common laboratory instrument that uses this technique is a Fourier transform infrared (FTIR) spectrometer.

The infrared portion of the electromagnetic spectrum is usually divided into three regions; the near-, mid- and far- infrared, named for their relation to the visible spectrum. The higher-energy near-IR, approximately 14000–4000 cm⁻¹ (0.8–2.5 µm wavelength) can excite overtone or harmonic vibrations. The mid-infrared, approximately 4000–400 cm⁻¹ (2.5–25 µm) may be used to study the fundamental vibrations and associated rotational-vibrational structure. The far-infrared, approximately 400–10 cm⁻¹ (25–1000 µm), lying adjacent to the microwave region, has low energy and may be used for rotational spectroscopy. The names and classifications of these subregions are conventions, and are only loosely based on the relative molecular or electromagnetic properties.

Theory

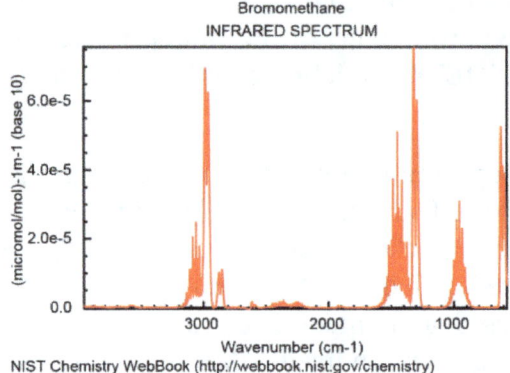

Sample of an IR spec. reading; this one is from bromomethane (CH_3Br), showing peaks around 3000, 1300, and 1000 cm⁻¹ (on the horizontal axis).

Infrared spectroscopy exploits the fact that molecules absorb frequencies that are characteristic of their structure. These absorptions are resonant frequencies, i.e. the frequency of the absorbed radiation matches the vibrational frequency. The energies are affected by the shape of the molecular potential energy surfaces, the masses of the atoms, and the associated vibronic coupling.

In particular, in the Born–Oppenheimer and harmonic approximations, i.e. when the molecular Hamiltonian corresponding to the electronic ground state can be approximated by a harmonic oscillator in the neighborhood of the equilibrium molecular geometry, the resonant frequencies are associated with the normal modes corresponding to the molecular electronic ground state potential energy surface. The resonant frequencies are also related to the strength of the bond and the mass of the atoms at either end of it. Thus, the frequency of the vibrations are associated with a particular normal mode of motion and a particular bond type.

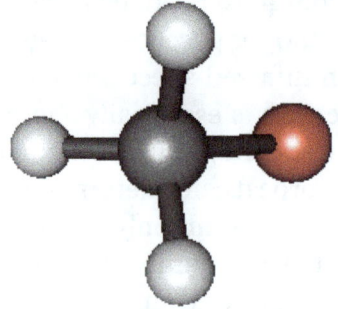

C–H bonds of bromomethane

Number of Vibrational Modes

In order for a vibrational mode in a sample to be "IR active", it must be associated with changes in the dipole moment. A permanent dipole is not necessary, as the rule requires only a change in dipole moment.

A molecule can vibrate in many ways, and each way is called a *vibrational mode*. For molecules with N number of atoms, linear molecules have $3N - 5$ degrees of vibrational modes, whereas nonlinear molecules have $3N - 6$ degrees of vibrational modes (also called vibrational degrees of freedom). As an example H_2O, a non-linear molecule, will have $3 \times 3 - 6 = 3$ degrees of vibrational freedom, or modes.

Simple diatomic molecules have only one bond and only one vibrational band. If the molecule is symmetrical, e.g. N_2, the band is not observed in the IR spectrum, but only in the Raman spectrum. Asymmetrical diatomic molecules, e.g. CO, absorb in the IR spectrum. More complex molecules have many bonds, and their vibrational spectra are correspondingly more complex, i.e. big molecules have many peaks in their IR spectra.

The atoms in a CH_2X_2 group, commonly found in organic compounds and where X can represent any other atom, can vibrate in nine different ways. Six of these vibrations involve only the CH_2 portion: symmetric and antisymmetric stretching, scissoring, rocking, wagging and twisting, as shown below. Structures that do not have the two additional X groups attached have fewer modes because some modes are defined by specific relationships to those other attached groups. For example, in water, the rocking, wagging, and twisting modes do not exist because these types of motions of the H represent simple rotation of the whole molecule rather than vibrations within it.

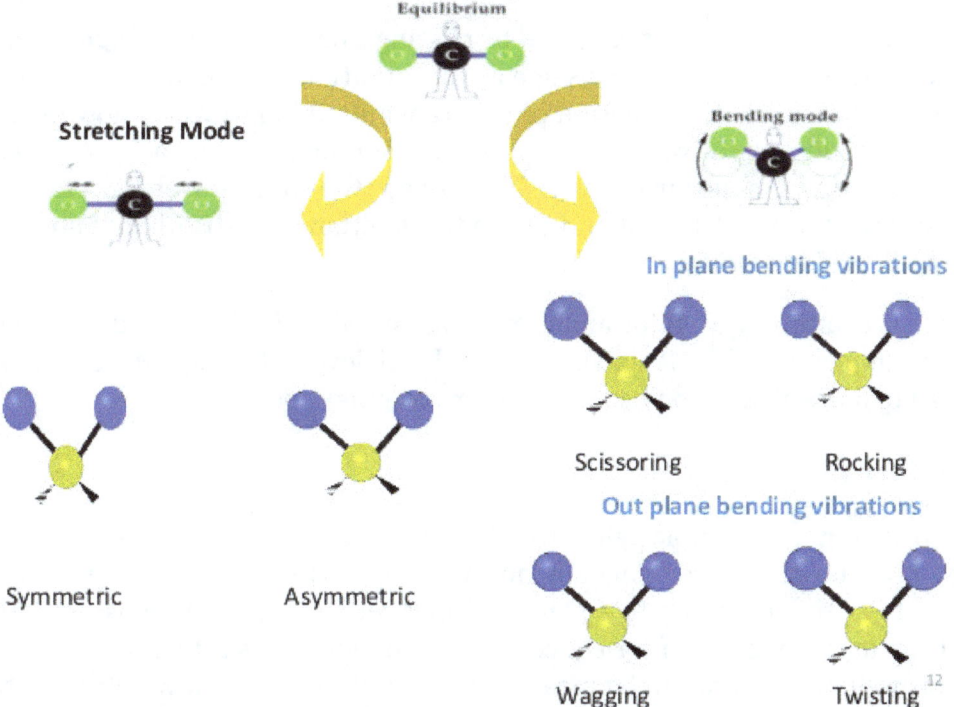

These figures do not represent the "recoil" of the C atoms, which, though necessarily present to balance the overall movements of the molecule, are much smaller than the movements of the lighter H atoms.

Special Effects

The simplest and most important IR bands arise from the "normal modes," the simplest distortions of the molecule. In some cases, "overtone bands" are observed. These bands arise from the absorption of a photon that leads to a doubly excited vibrational state. Such bands appear at approximately twice the energy of the normal mode. Some vibrations, so-called 'combination modes," involve more than one normal mode. The phenomenon of Fermi resonance can arise when two modes are similar in energy; Fermi resonance results in an unexpected shift in energy and intensity of the bands etc.

Practical IR Spectroscopy

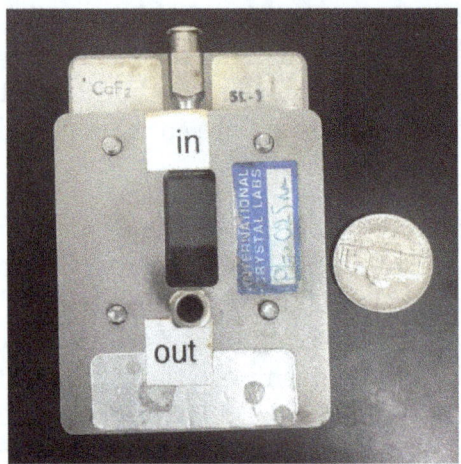

Typical IR solution cell. The windows are CaF_2.

The infrared spectrum of a sample is recorded by passing a beam of infrared light through the sample. When the frequency of the IR is the same as the vibrational frequency of a bond or collection of bonds, absorption occurs. Examination of the transmitted light reveals how much energy was absorbed at each frequency (or wavelength). This measurement can be achieved by scanning the wavelength range using a monochromator. Alternatively, the entire wavelength range is measured using a Fourier transform instrument and then a transmittance or absorbance spectrum is generated using a dedicated procedure.

This technique is commonly used for analyzing samples with covalent bonds. Simple spectra are obtained from samples with few IR active bonds and high levels of purity. More complex molecular structures lead to more absorption bands and more complex spectra.

Sample Preparation

Gaseous samples require a sample cell with a long pathlength to compensate for the diluteness. The pathlength of the sample cell depends on the concentration of the compound of interest. A simple glass tube with length of 5 to 10 cm equipped with infrared-transparent windows at the both ends of the tube can be used for concentrations down to several hundred ppm. Sample gas concentrations well below ppm can be measured with a White's cell in which the infrared light is guided with mirrors to travel through the gas. White's cells are available with optical pathlength starting from 0.5 m up to hundred meters.

Liquid samples can be sandwiched between two plates of a salt (commonly sodium chloride, or common salt, although a number of other salts such as potassium bromide or calcium fluoride are also used). The plates are transparent to the infrared light and do not introduce any lines onto the spectra.

Solid samples can be prepared in a variety of ways. One common method is to crush the sample with an oily mulling agent (usually mineral oil Nujol). A thin film of the mull is applied onto salt plates and measured. The second method is to grind a quantity of the sample with a specially purified salt (usually potassium bromide) finely (to remove scattering effects from large crystals). This powder mixture is then pressed in a mechanical press to form a translucent pellet through which the beam of the spectrometer can pass. A third technique is the "cast film" technique, which is used mainly for polymeric materials. The sample is first dissolved in a suitable, non hygroscopic solvent. A drop of this solution is deposited on surface of KBr or NaCl cell. The solution is then evaporated to dryness and the film formed on the cell is analysed directly. Care is important to ensure that the film is not too thick otherwise light cannot pass through. This technique is suitable for qualitative analysis. The final method is to use microtomy to cut a thin (20–100 µm) film from a solid sample. This is one of the most important ways of analysing failed plastic products for example because the integrity of the solid is preserved.

In photoacoustic spectroscopy the need for sample treatment is minimal. The sample, liquid or solid, is placed into the sample cup which is inserted into the photoacoustic cell which is then sealed for the measurement. The sample may be one solid piece, powder or basically in any form for the measurement. For example, a piece of rock can be inserted into the sample cup and the spectrum measured from it.

Comparing to a Reference

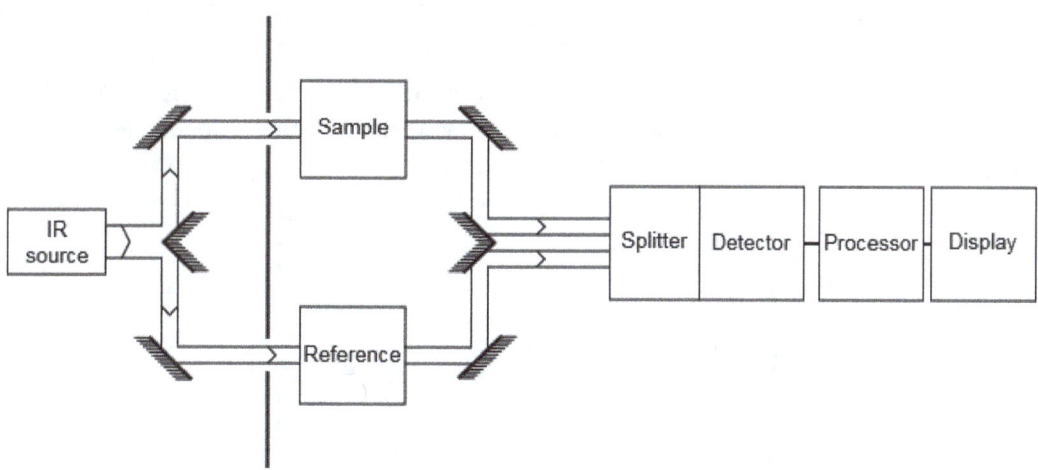

Schematics of a two-beam absorption spectrometer. A beam of infrared light is produced, passed through an interferometer (not shown), and then split into two separate beams. One is passed through the sample, the other passed through a reference. The beams are both reflected back towards a detector, however first they pass through a splitter, which quickly alternates which of the two beams enters the detector. The two signals are then compared and a printout is obtained. This "two-beam" setup gives accurate spectra even if the intensity of the light source drifts over time.

It is typical to record spectrum of both the sample and a "reference". This step controls for a number of variables, e.g. infrared detector, which may affect the spectrum. The reference measurement makes it possible to eliminate the instrument influence.

The appropriate "reference" depends on the measurement and its goal. The simplest reference measurement is to simply remove the sample (replacing it by air). However, sometimes a different reference is more useful. For example, if the sample is a dilute solute dissolved in water in a beaker, then a good reference measurement might be to measure pure water in the same beaker. Then the reference measurement would cancel out not only all the instrumental properties (like what light source is used), but also the light-absorbing and light-reflecting properties of the water and beaker, and the final result would just show the properties of the solute (at least approximately).

A common way to compare to a reference is sequentially: first measure the reference, then replace the reference by the sample and measure the sample. This technique is not perfectly reliable; if the infrared lamp is a bit brighter during the reference measurement, then a bit dimmer during the sample measurement, the measurement will be distorted. More elaborate methods, such as a "two-beam" setup, can correct for these types of effects to give very accurate results. The Standard addition method can be used to statistically cancel these errors.

Nevertheless, among different absorption based techniques which are used for gaseous species detection, Cavity ring-down spectroscopy (CRDS) can be used as a calibration free method. The fact that CRDS is based on the measurements of photon life-times (and not the laser intensity) makes it needless for any calibration and comparison with a reference.

FTIR

Fourier transform infrared (FTIR) spectroscopy is a measurement technique that allows one to record infrared spectra. Infrared light is guided through an interferometer and then through the sample (or vice versa). A moving mirror inside the apparatus alters the distribution of infrared light that passes through the interferometer. The signal directly recorded, called an "interferogram", represents light output as a function of mirror position. A data-processing technique called Fourier transform turns this raw data into the desired result (the sample's spectrum): Light output as a function of infrared wavelength (or equivalently, wavenumber). As described above, the sample's spectrum is always compared to a reference.

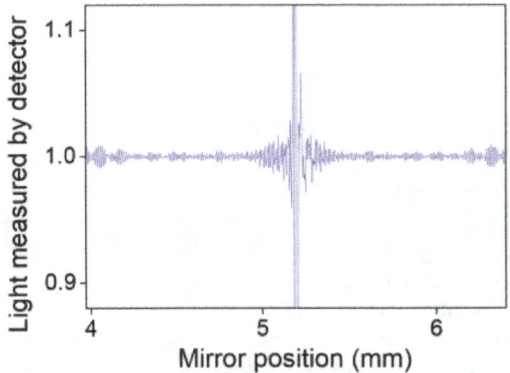

An interferogram from an FTIR measurement. The horizontal axis is the position of the mirror, and the vertical axis is the amount of light detected. This is the "raw data" which can be Fourier transformed to get the actual spectrum.

An alternate method for acquiring spectra is the "dispersive" or "scanning monochromator" method. In this approach, the sample is irradiated sequentially with various single wavelengths. The dispersive method is more common in UV-Vis spectroscopy, but is less practical in the infrared than the FTIR method. One reason that FTIR is favored is called "Fellgett's advantage" or the "multiplex advantage": The information at all frequencies is collected simultaneously, improving both speed and signal-to-noise ratio. Another is called "Jacquinot's Throughput Advantage": A dispersive measurement requires detecting much lower light levels than an FTIR measurement. There are other advantages, as well as some disadvantages, but virtually all modern infrared spectrometers are FTIR instruments.

Absorption Bands

IR spectroscopy is often used to identify structures because functional groups give rise to characteristic bands both in terms of intensity and position (frequency). The positions of these bands are summarized in correlation tables as shown below.

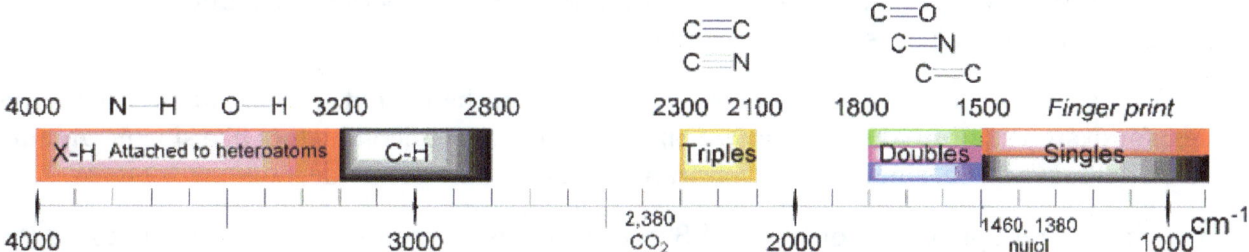

Wavenumbers listed in cm^{-1}.

Badger's Rule

For many kinds of samples, the assignments are known, i.e. which bond deformation(s) are associated with which frequency. In such cases further information can be gleaned about the strength on a bond, relying on the empirical guideline called Badger's Rule. Originally published by Richard Badger in 1934, this rule states that the strength of a bond correlates with the frequency of its vibrational mode. That is, increase in bond strength leads to corresponding frequency increase and vice versa.

Uses and Applications

Infrared spectroscopy is a simple and reliable technique widely used in both organic and inorganic chemistry, in research and industry. It is used in quality control, dynamic measurement, and monitoring applications such as the long-term unattended measurement of CO_2 concentrations in greenhouses and growth chambers by infrared gas analyzers.

It is also used in forensic analysis in both criminal and civil cases, for example in identifying polymer degradation. It can be used in determining the blood alcohol content of a suspected drunk driver.

IR-spectroscopy has been successfully used in analysis and identification of pigments in paintings and other art objects such as illuminated manuscripts.

A useful way of analyzing solid samples without the need for cutting samples uses ATR or attenuated total reflectance spectroscopy. Using this approach, samples are pressed against the face of a single crystal. The infrared radiation passes through the crystal and only interacts with the sample at the interface between the two materials.

With increasing technology in computer filtering and manipulation of the results, samples in solution can now be measured accurately (water produces a broad absorbance across the range of interest, and thus renders the spectra unreadable without this computer treatment).

Some instruments will also automatically tell you what substance is being measured from a store of thousands of reference spectra held in storage.

Infrared spectroscopy is also useful in measuring the degree of polymerization in polymer manufacture. Changes in the character or quantity of a particular bond are assessed by measuring at a specific frequency over time. Modern research instruments can take infrared measurements across the range of interest as frequently as 32 times a second. This can be done whilst simultaneous measurements are made using other techniques. This makes the observations of chemical reactions and processes quicker and more accurate.

Infrared spectroscopy has also been successfully utilized in the field of semiconductor microelectronics: for example, infrared spectroscopy can be applied to semiconductors like silicon, gallium arsenide, gallium nitride, zinc selenide, amorphous silicon, silicon nitride, etc.

Another important application of Infrared Spectroscopy is in the food industry to measure the concentration of various compounds in different food products.

The instruments are now small, and can be transported, even for use in field trials.

Infrared Spectroscopy is also used in gas leak detection devices such as the DP-IR and EyeC-GAs. These devices detect hydrocarbon gas leaks in the transportation of natural gas and crude oil.

In February 2014, NASA announced a greatly upgraded database, based on IR spectroscopy, for tracking polycyclic aromatic hydrocarbons (PAHs) in the universe. According to scientists, more than 20% of the carbon in the universe may be associated with PAHs, possible starting materials for the formation of life. PAHs seem to have been formed shortly after the Big Bang, are widespread throughout the universe, and are associated with new stars and exoplanets.

Isotope Effects

The different isotopes in a particular species may exhibit different fine details in infrared spectroscopy. For example, the O–O stretching frequency (in reciprocal centimeters) of oxyhemocyanin is experimentally determined to be 832 and 788 cm^{-1} for $v(^{16}O–^{16}O)$ and $v(^{18}O–^{18}O)$, respectively.

By considering the O–O bond as a spring, the wavenumber of absorbance, v can be calculated:

$$v = \frac{1}{2\pi c}\sqrt{\frac{k}{\mu}}$$

where k is the spring constant for the bond, c is the speed of light, and μ is the reduced mass of the A–B system:

$$\mu = \frac{m_A m_B}{m_A + m_B}$$

(m_i is the mass of atom i)

The reduced masses for ^{16}O–^{16}O and ^{18}O–^{18}O can be approximated as 8 and 9 respectively. Thus

$$\frac{v(^{16}O)}{v(^{18}O)} = \sqrt{\frac{9}{8}} \approx \frac{832}{788}.$$

Where v is the wavenumber; [wavenumber = frequency/(speed of light)]

The effect of isotopes, both on the vibration and the decay dynamics, has been found to be stronger than previously thought. In some systems, such as silicon and germanium, the decay of the anti-symmetric stretch mode of interstitial oxygen involves the symmetric stretch mode with a strong isotope dependence. For example, it was shown that for a natural silicon sample, the lifetime of the anti-symmetric vibration is 11.4 ps. When the isotope of one of the silicon atoms is increased to ^{29}Si, the lifetime increases to 19 ps. In similar manner, when the silicon atom is changed to ^{30}Si, the lifetime becomes 27 ps.

Two-dimensional IR

Two-dimensional infrared correlation spectroscopy analysis combines multiple samples of infrared spectra to reveal more complex properties. By extending the spectral information of a perturbed sample, spectral analysis is simplified and resolution is enhanced. The 2D synchronous and 2D asynchronous spectra represent a graphical overview of the spectral changes due to a perturbation (such as a changing concentration or changing temperature) as well as the relationship between the spectral changes at two different wavenumbers.

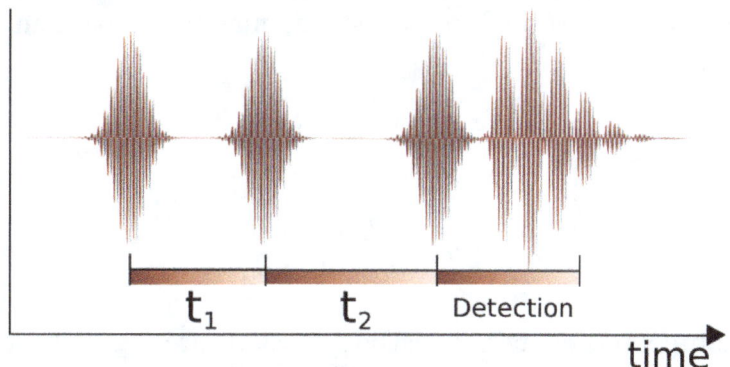

Pulse Sequence used to obtain a two-dimensional Fourier transform infrared spectrum. The time period T_1 is usually referred to as the coherence time and the second time period T_2 is known as the waiting time. The excitation frequency is obtained by Fourier transforming along the T_1 axis.

Nonlinear two-dimensional infrared spectroscopy is the infrared version of correlation spectroscopy. Nonlinear two-dimensional infrared spectroscopy is a technique that has become available with the development of femtosecond infrared laser pulses. In this experiment, first a set of pump pulses is applied to the sample. This is followed by a waiting time during which the system is allowed to relax. The typical waiting time lasts from zero to several picoseconds, and the duration can be controlled with a resolution of tens of femtoseconds. A probe pulse is then applied, resulting in the emission of a signal from the sample. The nonlinear two-dimensional infrared spectrum is a two-dimensional correlation plot of the frequency ω_1 that was excited by the initial pump pulses and the frequency ω_3 excited by the probe pulse after the waiting time. This allows the observation of coupling between different vibrational modes; because of its extremely fine time resolution, it can be used to monitor molecular dynamics on a picosecond timescale. It is still a largely unexplored technique and is becoming increasingly popular for fundamental research.

As with two-dimensional nuclear magnetic resonance (2DNMR) spectroscopy, this technique spreads the spectrum in two dimensions and allows for the observation of cross peaks that contain information on the coupling between different modes. In contrast to 2DNMR, nonlinear two-dimensional infrared spectroscopy also involves the excitation to overtones. These excitations result in excited state absorption peaks located below the diagonal and cross peaks. In 2DNMR, two distinct techniques, COSY and NOESY, are frequently used. The cross peaks in the first are related to the scalar coupling, while in the latter they are related to the spin transfer between different nuclei. In nonlinear two-dimensional infrared spectroscopy, analogs have been drawn to these 2DNMR techniques. Nonlinear two-dimensional infrared spectroscopy with zero waiting time corresponds to COSY, and nonlinear two-dimensional infrared spectroscopy with finite waiting time allowing vibrational population transfer corresponds to NOESY. The COSY variant of nonlinear two-dimensional infrared spectroscopy has been used for determination of the secondary structure content of proteins.

IR Spectroscopy

Qualitative to semi-quantitative analysis of organometallic compounds using IR spectroscopy are performed whenever possible. In general the signature stretching vibrations for chemical bonds are more conveniently looked at in these studies. The frequency (v) of a stretching vibration of a covalent bond is directly proportional to the strength of the bond, usually given by the force constant (k) and inversely proportional to the reduced mass of the system, which relates to the masses of the individual atoms.

$$v = \frac{1}{2\pi C}\left\{ \sqrt{\frac{k}{m_r}} \right\}$$

$$m_r = \frac{m_1 m_2}{m_1 + m_2}$$

The organometallic compounds containing carbonyl groups are regularly studied using IR spectroscopy, and in which the CO peaks appear in the range between 2100–1700 cm^{-1} as distinctly intense peaks.

Crystallography

The solid state structure elucidation using single crystal diffraction studies are extremely useful techniques for the characterization of the organometallic compounds and for which the X-ray diffraction and neutron diffraction studies are often undertaken. As these methods give a three dimensional structural rendition at a molecular level, they are of significant importance among the various available characterization methods. The X-ray diffraction technique is founded on Bragg's law that explains the diffraction pattern arising out of a repetitive arrangement of the atoms located at the crystal lattices.

$$2d \sin \theta = n\lambda$$

A major limitation of the X-ray diffraction is that the technique is not sensitive enough to detect the hydrogen atoms, which appear as weak peaks as opposed to intense peaks arising out of the more electron rich metal atoms, and hence are not very useful for metal hydride compounds. Neutron diffraction studies can detect hydrogens more accurately and thus are good for the analysis of the metal hydride complexes.

Molecular Structure and Bonding

The chemical properties of the molecules can be directly correlated to their electronic structures. In this chapter attempts are being made to give an overall view of how bonding concept evolved starting from Lewis approach to the development of molecular orbital theory.

Lewis Structures

Lewis proposed that when two atoms come close to each other to establish a bond by sharing an electron pair, a covalent bond will be established. One pair of electrons would give a single bond X—Y; two or three pairs of electrons would leads to the formation of double (X=Y) and triple bonds (X≡Y), respectively. The pairs of valence electrons that are not utilized in bonding are called lone pairs of electrons or simply lone pairs. The lone pair of electrons does not participate in bonding; however, they do influence the shape and the geometry of the molecule and their chemical properties as well.

Lewis introduced octet rule which states that each atom shares its valence electrons with neighboring atoms to have a total of eight (s^2p^6) electrons in its valence shell to have noble-gas configuration. An exception to this rule is hydrogen as it can have only two valence electrons in its only shell, 1s.

By simply counting the number of valence electrons present on the central atom and its neighbors, Lewis structures can be written in just three easy steps.

1. Consider the valence electrons of all participating atoms; add an electron for each negative charge and subtract one electron for each positive charge.

2. Identify the central atom and write the symbols of the atoms around central atom. In majority of polyatomic molecules, the least electronegative one will be the central atom with an exception of hydrides, for example, H_2O, NH_3 or H_2S.

3. Distribute the electron pairs throughout the molecule to satisfy the octet of all atoms pres-

ent in the molecule starting from the most electronegative one. Each pair of singly bonded atoms requires one pair of electrons.

4. Each bonding pair should be represented by a single bond and the net charge is assumed to be possessed by the ion (cation or anion) as a whole and not by an individual atom.

For some molecules, Lewis dot structure differs from the experimentally determined structural observations. For example, in acetate ion both the C—O bonds are identical as per X-ray structure determination but the prediction by the Lewis structure is incorrect. The reason is due to resonance.

Limitations of Lewis Model

Molecules with odd number of electrons can never satisfy the octet rule.

Example: NO.

Some atoms with fewer valence electrons can never complete octet without formal charges.

A central atom can have more than 8 electrons. Example: SF_6

Lewis model does not explain paramagnetic nature of O_2.

Geometry and molecular shapes cannot be explained by Lewis model.

Worked Examples

Example 1 : ClO_2^-

Solution :

ClO_2^- Total number of electrons $7 + 2 \times 6 + 1 = 20 = 10$ pairs

Identify the central atom and connect the peripheral atoms with a pair of electrons as two dots and count the remaining electron pairs $O:Cl:O$ $20-4 = 16$ electrons left (8 pairs)

Complete the octet of oxygen atoms and count the remaining electron pairs $:\overset{..}{O}:Cl:\overset{..}{O}:$ $16-12 = 4$ electrons

Complete the octet of chlorine atom, $:\overset{..}{O}:\overset{..}{Cl}:\overset{..}{O}:$ Since all the electrons are utilized and octet is satisfied, no need of any multiple bonds.

The structure is $\left[:\overset{..}{O}:\overset{..}{Cl}:\overset{..}{O}: \right]^-$ or $\left[|\overline{\underline{O}}:\overline{\underline{Cl}}:\overline{\underline{O}}| \right]^-$ or $\left[:\overset{..}{O}\!\!-\!\!\overset{.}{\underset{.}{Cl}}\!\!-\!\!\overset{..}{O}: \right]^-$

Lewis Acids and Bases

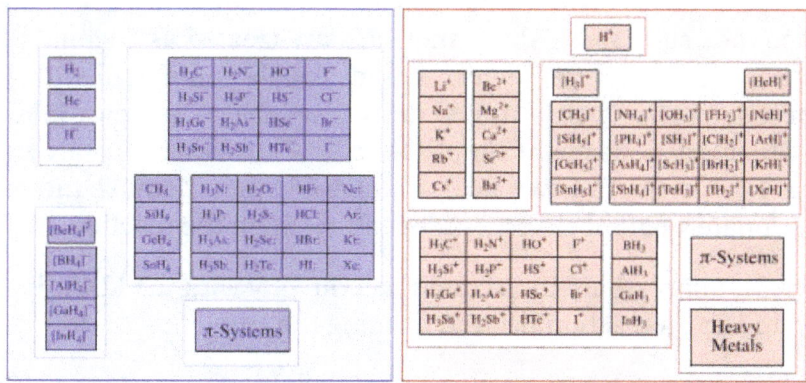

Diagram of some Lewis bases and acids

A Lewis acid is a chemical species that reacts with a Lewis base to form a Lewis adduct. A Lewis base, then, is any species that donates a pair of electrons to a Lewis acid to form a Lewis adduct. For example, OH⁻ and NH_3 are Lewis bases, because they can donate a lone pair of electrons. In the adduct, the Lewis acid and base share an electron pair furnished by the Lewis base. Usually the terms Lewis acid and Lewis base are defined within the context of a specific chemical reaction. For example, in the reaction of Me_3B and NH_3 to give Me_3BNH_3, Me_3B acts as a Lewis acid, and NH_3 acts as a Lewis base. Me_3BNH_3 is the Lewis adduct. The terminology refers to the contributions of Gilbert N. Lewis.

Depicting Adducts

In many cases, the interaction between the Lewis base and Lewis acid in a complex is indicated by an arrow indicating the Lewis base donating electrons toward the Lewis acid using the notation of a dative bond—for example, $Me_3B{\leftarrow}NH_3$. Some sources indicate the Lewis base with a pair of dots (the explicit electrons being donated), which allows consistent representation of the transition from the base itself to the complex with the acid:

$$Me_3B + \; :NH_3 \rightarrow Me_3B{:}NH_3$$

A center dot may also be used to represent a Lewis adduct, such as $Me_3B{\bullet}NH_3$. Another example is boron trifluoride etherate, $BF_3{\bullet}Et_2O$. The center dot is also used to represent hydrate coordination in various crystals, as in $MgSO_4{\bullet}7H_2O$ for hydrated magnesium sulfate. In general, however, the donor–acceptor bond is viewed as simply somewhere along a continuum between idealized covalent bonding and ionic bonding.

Examples

Major structural changes accompany binding of the Lewis base
to the coordinatively unsaturated, planar Lewis acid BF_3.

Classically, the term "Lewis acid" is restricted to trigonal planar species with an empty p orbital, such as BR_3 where R can be an organic substituent or a halide. For the purposes of discussion, even complex compounds such as $Et_3Al_2Cl_3$ and $AlCl_3$ are treated as trigonal planar Lewis acids. Metal ions such as Na^+, Mg^{2+}, and Ce^{3+}, which are invariably complexed with additional ligands, are often sources of coordinatively unsaturated derivatives that form Lewis adducts upon reaction with a Lewis base. Other reactions might simply be referred to as "acid-catalyzed" reactions. Some compounds, such as H_2O, are both Lewis acids and Lewis bases, because they can either accept a pair of electrons or donate a pair of electrons, depending upon the reaction.

Lewis acids are diverse. Simplest are those that react directly with the Lewis base. But more common are those that undergo a reaction prior to forming the adduct.

- Examples of Lewis acids based on the general definition of electron pair acceptor include:

    o   the proton (H^+) and acidic compounds onium ions, such as NH_4^+ and H_3O^+

    o   metal cations, such as Li^+ and Mg^{2+}, often as their aquo or ether complexes,

    o   trigonal planar species, such as BF_3 and carbocations H_3C^+

    o   pentahalides of phosphorus, arsenic, and antimony

    o   electron poor π-systems, such as enones and tetracyanoethylenes.

Again, the description of a Lewis acid is often used loosely. For example, in solution, bare protons do not exist.

Simple Lewis acids

The most studied examples of such Lewis acids are the boron trihalides and organoboranes, but other compounds exhibit this behavior:

$$BF_3 + F^- \rightarrow BF_4^-$$

In this adduct, all four fluoride centres (or more accurately, ligands) are equivalent.

$$BF_3 + OMe_2 \rightarrow BF_3OMe_2$$

Both BF_4^- and BF_3OMe_2 are Lewis base adducts of boron trifluoride.

In many cases, the adducts violate the octet rule, such as the triiodide anion:

$$I_2 + I^- \rightarrow I_3^-$$

The variability of the colors of iodine solutions reflects the variable abilities of the solvent to form adducts with the Lewis acid I_2.

In some cases, the Lewis acids are capable of binding two Lewis bases, a famous example being the formation of hexafluorosilicate:

$$SiF_4 + 2\ F^- \rightarrow SiF_6^{2-}$$

Complex Lewis acids

Most compounds considered to be Lewis acids require an activation step prior to formation of the adduct with the Lewis base. Well known cases are the aluminium trihalides, which are widely viewed as Lewis acids. Aluminium trihalides, unlike the boron trihalides, do not exist in the form AlX_3, but as aggregates and polymers that must be degraded by the Lewis base. A simpler case is the formation of adducts of borane. Monomeric BH_3 does not exist appreciably, so the adducts of borane are generated by degradation of diborane:

$$B_2H_6 + 2\,H^- \rightarrow 2\,BH_4^-$$

In this case, an intermediate $B_2H_7^-$ can be isolated.

Many metal complexes serve as Lewis acids, but usually only after dissociating a more weakly bound Lewis base, often water.

$$[Mg(H_2O)_6]^{2+} + 6\,NH_3 \rightarrow [Mg(NH_3)_6]^{2+} + 6\,H_2O$$

H+ as Lewis Acid

The proton (H^+) is one of the strongest but is also one of the most complicated Lewis acids. It is convention to ignore the fact that a proton is heavily solvated (bound to solvent). With this simplification in mind, acid-base reactions can be viewed as the formation of adducts:

- $H^+ + NH_3 \rightarrow NH_4^+$

- $H^+ + OH^- \rightarrow H_2O$

Applications of Lewis Acids

A typical example of a Lewis acid in action is in the Friedel–Crafts alkylation reaction. The key step is the acceptance by $AlCl_3$ of a chloride ion lone-pair, forming $AlCl_4^-$ and creating the strongly acidic, that is, electrophilic, carbonium ion.

$$RCl + AlCl_3 \rightarrow R^+ + AlCl_4^-$$

Lewis Bases

A Lewis base is an atomic or molecular species where the highest occupied molecular orbital (HOMO) is highly localized. Typical Lewis bases are conventional amines such as ammonia and alkyl amines. Other common Lewis bases include pyridine and its derivatives. Some of the main classes of Lewis bases are:

- amines of the formula $NH_{3-x}R_x$ where R = alkyl or aryl. Related to these are pyridine and its derivatives.

- phosphines of the formula $PR_{3-x}A_x$, where R = alkyl, A = aryl.

- compounds of O, S, Se and Te in oxidation state 2, including water, ethers, ketones.

The most common Lewis bases are anions. The strength of Lewis basicity correlates with the pK_a of the parent acid: acids with high pK_a's give good Lewis bases. As usual, a weaker acid has a stronger conjugate base.

- Examples of Lewis bases based on the general definition of electron pair donor include:

  o simple anions, such as H^- and F^-.

  o other lone-pair-containing species, such as H_2O, NH_3, HO^-, and CH_3^-.

  o complex anions, such as sulfate.

  o electron rich π-system Lewis bases, such as ethyne, ethene, and benzene.

The strength of Lewis bases have been evaluated for various Lewis acids, such as I_2, $SbCl_5$, and BF_3.

Heats of binding of various bases to BF_3		
Lewis base	donor atom	Enthalpy of Complexation (kJ/ mol)
Et_3N	N	135
quinuclidine	N	150
pyridine	N	128
Acetonitrile	N	60
Et_2O	O	78.8
THF	O	90.4
acetone	O	76.0
EtOAc	O	75.5
DMA	O	112
DMSO	O	105
Tetrahydrothiophene	S	51.6
Trimethylphosphine	P	97.3

Applications of Lewis Bases

Nearly all electron pair donors that form compounds by binding transition elements can be viewed as a collections of the Lewis bases – or ligands. Thus a large application of Lewis bases is to modify the activity and selectivity of metal catalysts. Chiral Lewis bases thus confer chirality on a catalyst, enabling asymmetric catalysis, which is useful for the production of pharmaceuticals.

Many Lewis bases are "multidentate," that is they can form several bonds to the Lewis acid. These multidentate Lewis acids are called chelating agents.

Hard and soft classification

Lewis acids and bases are commonly classified according to their hardness or softness. In this context hard implies small and nonpolarizable and soft indicates larger atoms that are more polarizable.

- typical hard acids: H^+, alkali/alkaline earth metal cations, boranes, Zn^{2+}

- typical soft acids: Ag^+, $Mo(0)$, $Ni(0)$, Pt^{2+}

- typical hard bases: ammonia and amines, water, carboxylates, fluoride and chloride

- typical soft bases: organophosphines, thioethers, carbon monoxide, iodide

For example, an amine will displace phosphine from the adduct with the acid BF_3. In the same way, bases could be classified. For example, bases donating a lone pair from an oxygen atom are harder than bases donating through a nitrogen atom. Although the classification was never quantified it proved to be very useful in predicting the strength of adduct formation, using the key concepts that hard acid — hard base and soft acid — soft base interactions are stronger than hard acid — soft base or soft acid — hard base interactions. Later investigation of the thermodynamics of the interaction suggested that hard—hard interactions are enthalpy favored, whereas soft—soft are entropy favored.

History

The concept originated with Gilbert N. Lewis who studied chemical bonding. In 1923, Lewis wrote *An acid substance is one which can employ an electron lone pair from another molecule in completing the stable group of one of its own atoms.* The Brønsted–Lowry acid–base theory was published in the same year. The two theories are distinct but complementary. A Lewis base is also a Brønsted–Lowry base, but a Lewis acid doesn't need to be a Brønsted–Lowry acid. The classification into hard and soft acids and bases (HSAB theory) followed in 1963. The strength of Lewis acid-base interactions, as measured by the standard enthalpy of formation of an adduct can be predicted by the Drago–Wayland two-parameter equation.

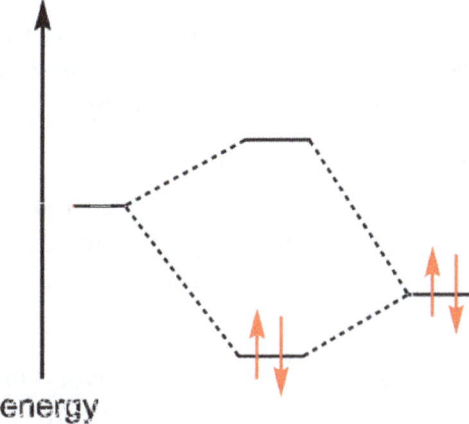

MO diagram depicting the formation of a dative covalent bond between two atoms.

Reformulation of Lewis Theory

Lewis had suggested in 1916 that two atoms are held together in a chemical bond by sharing a pair of electrons. When each atom contributed one electron to the bond it was called a covalent bond. When both electrons come from one of the atoms it was called a dative covalent bond or coordinate bond. The distinction is not very clear-cut. For example, in the formation of an ammonium ion from ammonia and hydrogen the ammonia molecule donates a pair of electrons to the proton; the identity of the electrons is lost in the ammonium ion that is formed. Nevertheless, Lewis suggested that an electron-pair donor be classified as a base and an electron-pair acceptor be classified as acid.

A more modern definition of a Lewis acid is an atomic or molecular species with a localized empty atomic or molecular orbital of low energy. This lowest energy molecular orbital (LUMO) can accommodate a pair of electrons.

Comparison with Brønsted–Lowry Theory

A Lewis base is often a Brønsted–Lowry base as it can donate a pair of electrons to H^+; the proton is a Lewis acid as it can accept a pair of electrons. The conjugate base of a Brønsted–Lowry acid is also a Lewis base as loss of H^+ from the acid leaves those electrons which were used for the A—H bond as a lone pair on the conjugate base. However, a Lewis base can be very difficult to protonate, yet still react with a Lewis acid. For example, carbon monoxide is a very weak Brønsted–Lowry base but it forms a strong adduct with BF_3.

In another comparison of Lewis and Brønsted–Lowry acidity by Brown and Kanner, 2,6-di-t-butylpyridine reacts to form the hydrochloride salt with HCl but does not react with BF_3. This example demonstrates that steric factors, in addition to electron configuration factors, play a role in determining the strength of the interaction between the bulky di-t-butylpyridine and tiny proton.

A Brønsted–Lowry acid is a proton donor, not an electron-pair acceptor.

Lewis Structure

Lewis structures (also known as Lewis dot diagrams, Lewis dot formulas,Lewis dot structures, and electron dot structures) are diagrams that show the bonding between atoms of a molecule and the lone pairs of electrons that may exist in the molecule. A Lewis structure can be drawn for any covalently bonded molecule, as well as coordination compounds. The Lewis structure was named after Gilbert N. Lewis, who introduced it in his 1916 article *The Atom and the Molecule*. Lewis structures extend the concept of the electron dot diagram by adding lines between atoms to represent shared pairs in a chemical bond.

Lewis structures show each atom and its position in the structure of the molecule using its chemical symbol. Lines are drawn between atoms that are bonded to one another (pairs of dots can be used instead of lines). Excess electrons that form lone pairs are represented as pairs of dots, and are placed next to the atoms.

Although main group elements of the second period and beyond usually react by gaining, losing, or sharing electrons until they have achieved a valence shell electron configuration with a full octet of (8) electrons, other elements obey different rules. Hydrogen (H) can only form bonds which share just two electrons, while transition metals often conform to a duodectet (12) rule (e.g., compounds such as the permanganate ion).

Construction

Counting Electrons

The total number of electrons represented in a Lewis structure is equal to the sum of the numbers of valence electrons on each individual atom. Non-valence electrons are not represented in Lewis structures.

Once the total number of available electrons has been determined, electrons must be placed into the structure. They should be placed initially as lone pairs: one pair of dots for each pair of electrons available. Lone pairs should initially be placed on outer atoms (other than hydrogen) until each outer atom has *eight* electrons in bonding pairs and lone pairs; extra lone pairs may then be placed on the central atom. When in doubt, lone pairs should be placed on more electronegative atoms first.

Once all lone pairs are placed, atoms—especially the central atoms—may not have an octet of electrons. In this case, the atoms must form a double bond; a lone pair of electrons is moved to form a second bond between the two atoms. As the bonding pair is shared between the two atoms, the atom that originally had the lone pair still has an octet; the other atom now has two more electrons in its valence shell.

Lewis structures for polyatomic ions may be drawn by the same method. When counting electrons, negative ions should have extra electrons placed in their Lewis structures; positive ions should have fewer electrons than an uncharged molecule.

When the Lewis structure of an ion is written, the entire structure is placed in brackets, and the charge is written as a superscript on the upper right, outside the brackets.

A simpler method has been proposed for constructing Lewis structures, eliminating the need for electron counting: the atoms are drawn showing the valence electrons; bonds are then formed by pairing up valence electrons of the atoms involved in the bond-making process, and anions and cations are formed by adding or removing electrons to/from the appropriate atoms.

A trick is to count up valence electrons, then count up the number of electrons needed to complete the octet rule (or with hydrogen just 2 electrons), then take the difference of these two numbers and the answer is the number of electrons that make up the bonds. The rest of the electrons just go to fill all the other atoms' octets.

Another simple and general procedure to write Lewis structures and resonance forms has been proposed.

Formal Charge

In terms of Lewis structures, formal charge is used in the description, comparison, and assessment of likely topological and resonance structures by determining the apparent electronic charge of each atom within, based upon its electron dot structure, assuming exclusive covalency or non-polar bonding. It has uses in determining possible electron re-configuration when referring to reaction mechanisms, and often results in the same sign as the partial charge of the atom, with exceptions. In general, the formal charge of an atom can be calculated using the following formula, assuming non-standard definitions for the markup used:

$$C_f = N_v - U_e - \frac{B_n}{2}$$

where:

- C_f is the formal charge.

- N_v represents the number of valence electrons in a free atom of the element.

- U_e represents the number of unshared electrons on the atom.

- B_n represents the total number of electrons in bonds the atom has with another.

The formal charge of an atom is computed as the difference between the number of valence electrons that a neutral atom would have and the number of electrons that belong to it in the Lewis structure. Electrons in covalent bonds are split equally between the atoms involved in the bond. The total of the formal charges on an ion should be equal to the charge on the ion, and the total of the formal charges on a neutral molecule should be equal to zero.

Resonance

For some molecules and ions, it is difficult to determine which lone pairs should be moved to form double or triple bonds, and two or more different *resonance* structures may be written for the same molecule or ion. In such cases it is usual to write all of them with two-way arrows in between. This is sometimes the case when multiple atoms of the same type surround the central atom, and is especially common for polyatomic ions.

When this situation occurs, the molecule's Lewis structure is said to be a resonance structure, and the molecule exists as a resonance hybrid. Each of the different possibilities is superimposed on the others, and the molecule is considered to have a Lewis structure equivalent to some combination of these states.

The nitrate ion (NO_3^-), for instance, must form a double bond between nitrogen and one of the oxygen's to satisfy the octet rule for nitrogen. However, because the molecule is symmetrical, it does not matter *which* of the oxygen's forms the double bond. In this case, there are three possible resonance structures. Expressing resonance when drawing Lewis structures may be done either by drawing each of the possible resonance forms and placing double-headed arrows between them or by using dashed lines to represent the partial bonds (although the latter is a good representation of the resonance hybrid which is not, formally speaking, a Lewis structure).

When comparing resonance structures for the same molecule, usually those with the fewest formal charges contribute more to the overall resonance hybrid. When formal charges are necessary, resonance structures that have negative charges on the more electronegative elements and positive charges on the less electronegative elements are favored.

Single bonds can also be moved in the same way to create resonance structures for hypervalent molecules such as sulfur hexafluoride, which is the correct description according to quantum chemical calculations instead of the common expanded octet model.

The resonance structure should not be interpreted to indicate that the molecule switches between forms, but that the molecule acts as the average of multiple forms.

Example

The formula of the nitrite ion is NO_2^-.

1. Nitrogen is the less electronegative atom of the two, so it is the central atom by multiple criteria.

2. Count valence electrons. Nitrogen has 5 valence electrons; each oxygen has 6, for a total of $(6 \times 2) + 5 = 17$. The ion has a charge of -1, which indicates an extra electron, so the total number of electrons is 18.

3. Place lone pairs. Each oxygen must be bonded to the nitrogen, which uses four electrons—two in each bond. The 14 remaining electrons should initially be placed as 7 lone pairs. Each oxygen may take a maximum of 3 lone pairs, giving each oxygen 8 electrons including the bonding pair. The seventh lone pair must be placed on the nitrogen atom.

4. Satisfy the octet rule. Both oxygen atoms currently have 8 electrons assigned to them. The nitrogen atom has only 6 electrons assigned to it. One of the lone pairs on an oxygen atom must form a double bond, but either atom will work equally well. Therefore, there is a resonance structure.

5. Tie up loose ends. Two Lewis structures must be drawn: Each structure has one of the two oxygen atoms double-bonded to the nitrogen atom. The second oxygen atom in each structure will be single-bonded to the nitrogen atom. Place brackets around each structure, and add the charge ($-$) to the upper right outside the brackets. Draw a double-headed arrow between the two resonance forms.

Alternative Formats

A skeletal diagram of butane

Chemical structures may be written in more compact forms, particularly when showing organic molecules. In condensed structural formulas, many or even all of the covalent bonds may be left out, with subscripts indicating the number of identical groups attached to a particular atom. Another shorthand structural diagram is the skeletal formula (also known as a bond-line formula or carbon skeleton diagram). In a skeletal formula, carbon atoms are not signified by the symbol C but by the vertices of the lines. Hydrogen atoms bonded to carbon are not shown—they can be inferred by counting the number of bonds to a particular carbon atom—each carbon is assumed to have four bonds in total, so any bonds not shown are, by implication, to hydrogen atoms.

Other diagrams may be more complex than Lewis structures, showing bonds in 3D using various forms such as space-filling diagrams.

VSEPR Theory and its Utility

The Valence Shell Electron Pair Repulsion Theory (VSEPR)

VSEPR theory is an improved and extension of Lewis model but predicts the shapes of polyatomic molecules. This model was first suggested by Nevil Sidgwick and Herbet Powell in 1940 and later improved by Ronald Gillespie and Ronald Nyholm.

Prediction of molecular shapes and geometries was made easy by this model through the following simple steps.

- Draw the Lewis structure.

- Count the total number of bonds and lone pairs around the central atom. (Each single bond would involve one pair of electrons).

- Arrange the bonding pairs and lone pairs in one of the standard geometries to minimize the electron-electron repulsion.

  o   Lone pair electrons stay closer to the nucleus and also they spread out over a larger space than bond pairs and hence large angles between lone pairs.

  o   The repulsion follows the order LP—LP > LP—BP > BP—BP.

  - Multiple bonds should be considered as a single bonding region.

Steric Numbers:

Another term called steric number is often used in VSEPR theory.

Steric number (SN) = No. of attached atom + No. of lone pairs. Since the lone pair—lone pair repulsions are maximum, the most stable geometry can be obtained by maximizing the distance between steric numbers on the central atom.

Molecular shapes are eventually determined by two parameters: Bond distance, separation between the nuclei of two bonded atoms in a straight line and the bond angle, the angle between any two bonds containing a common atom.

While mentioning the molecular shapes lone pairs may be ignored, however, while defining the geometry both the lone pairs and bond pairs should be considered.

For example: in water molecule the central oxygen atom is in tetrahedral environment with two lone pairs and two O—H bonds (or two bond pairs). The shape of the water molecule is therefore bent (two lone pairs are ignored).

Similarly, in ammonia, the nitrogen atom is in tetrahedral environment with three bonded pairs (three N—H bonds) and one lone pair. The shape of NH_3 molecule is pyramidal.

Predicting the Molecular Geometries

To begin with, draw the Lewis structure.

Count the number of bonding pairs and lone pairs around the central atom.

Arrange the bonding pairs and the lone pairs in one of the standard geometries thereby minimizing electron—electron repulsion.

Multiple bonds count as a single bonding region.

What is Bent's Rule:

More electronegative substituents 'prefer' hybrid orbitals having less s-character, and more electropositive substituents 'prefer' orbitals having more s-character.

The bond angles in CH_4, CF_4 and CH_2F_2 can be explained using Bent's rule. While a carbon in CH4 and CF_4 uses four identical sp³ hybrids for bonding, in CH_2F_2 the hybrids used are not identical.

The C-F bonds are formed from $^{sp3+x}$ hybrids, with slightly more p-character and less s-character than an sp³ hybrid, and the hydrogen are bonded by sp^{3-x} hybrids, with slightly less p-character and slightly more s-character. Increasing the amount of p-character in the C-F bonds decreases the F-C-F bond angle, because for bonding by pure p-orbitals the bond angle would be decreased to 90°.

Molecular Shapes Determined by VSEPR Theory

Molecule	Steric Number (Number Electron Pairs) (SN)	Geometry	Example
MA_2	2	Linear	$BeCl_2$
MA_3	3	Trigonal planar	BF_3
MA_4	4	Tetrahedral	SiF_4
MA_5	5	Trigonal bipyramidal	PF_5
MA_6	6	Octahedral	SF_6
MA_7	7	Pentagonal bipyramidal	IF_7

Molecule	SN	Number of lone pairs	Geometry	shape	Example
MA_2	2	0	Linear		CO_2
MA_3	3 3	0 1	Trigonal planar	Trigonal planar angular	SO_3 SO_2
MA_4	4	0 1 2	Tetrahedral	Tetrahedral Trigonal pyramidal Angular	CH_4 NH_3 H_2O

MA$_5$	5	0 1 2 3	Trigonal bipyrami- dal	Trigonal bipyramidal Seesaw T-shaped linear	AsF$_5$ SF$_4$ ClF$_3$ XeF$_2$
MA$_6$	6	0 1 2	Octahedral	Octahedral Square pyramidal Square planar	SF$_6$ BrF$_5$ XeF$_4$

Orbital Hybridisation

In chemistry, hybridisation (or hybridization) is the concept of mixing atomic orbitals into new *hybrid orbitals* (with different energies, shapes, etc., than the component atomic orbitals) suitable for the pairing of electrons to form chemical bonds in valence bond theory. Hybrid orbitals are very useful in the explanation of molecular geometry and atomic bonding properties. Although sometimes taught together with the valence shell electron-pair repulsion (VSEPR) theory, valence bond and hybridisation are in fact not related to the VSEPR model.

History

Chemist Linus Pauling first developed the hybridisation theory in 1931 in order to explain the structure of simple molecules such as methane (CH_4) using atomic orbitals. Pauling pointed out that a carbon atom forms four bonds by using one s and three p orbitals, so that "it might be inferred" that a carbon atom would form three bonds at right angles (using p orbitals) and a fourth weaker bond using the s orbital in some arbitrary direction. In reality however, methane has four bonds of equivalent strength separated by the tetrahedral bond angle of 109.5°. Pauling explained this by supposing that in the presence of four hydrogen atoms, the s and p orbitals form four equivalent combinations or *hybrid* orbitals, each denoted by sp³ to indicate its composition, which are directed along the four C-H bonds. This concept was developed for such simple chemical systems, but the approach was later applied more widely, and today it is considered an effective heuristic for rationalising the structures of organic compounds. It gives a simple orbital picture equivalent to Lewis structures. Hybridisation theory finds its use mainly in organic chemistry.

Atomic Orbitals

Orbitals are a model representation of the behaviour of electrons within molecules. In the case of simple hybridisation, this approximation is based on atomic orbitals, similar to those obtained for the hydrogen atom, the only neutral atom for which the Schrödinger equation can be solved exactly. In heavier atoms, such as carbon, nitrogen, and oxygen, the atomic orbitals used are the 2s and 2p orbitals, similar to excited state orbitals for hydrogen.

Overview

Hybrid orbitals are assumed to be mixtures of atomic orbitals, superimposed on each other in various proportions. For example, in methane, the C hybrid orbital which forms each carbon–hydrogen bond consists of 25% s character and 75% p character and is thus described as sp³ (read as *s-p-three*) hybridised. Quantum mechanics describes this hybrid as an sp³ wavefunction of the form $N(s + \sqrt{3}p\sigma)$, where N is a normalisation constant (here 1/2) and pσ is a p orbital directed along the C-H axis to form a sigma bond. The ratio of coefficients (denoted λ in general) is $\sqrt{3}$ in this example. Since the electron density associated with an orbital is proportional to the square of the wavefunction, the ratio of p-character to s-character is $\lambda^2 = 3$. The p character or the weight of the p component is $N^2\lambda^2 = 3/4$.

The amount of p character or s character, which is decided mainly by orbital hybridisation, can be used to reliably predict molecular properties such as acidity or basicity.

Two Possible Representations

Molecules with multiple bonds or multiple lone pairs can have orbitals represented in terms of sigma and pi symmetry or equivalent orbitals. The sigma and pi representation of Erich Hückel is the more common one compared to the equivalent orbital representation of Linus Pauling. The two have mathematically equivalent total many-electron wave functions, and are related by a unitary transformation of the set of occupied molecular orbitals.

Types of Hybridisation

sp3

Hybridisation describes the bonding atoms from an atom's point of view. For a tetrahedrally coordinated carbon (e.g., methane CH_4), the carbon should have 4 orbitals with the correct symmetry to bond to the 4 hydrogen atoms.

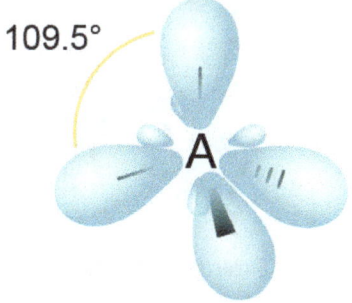

109.5°

Four sp³ orbitals.

Carbon's ground state configuration is $1s^2\ 2s^2\ 2p^2$ or more easily read:

C	↑↓	↑↓	↑	↑	
	1s	2s	2p	2p	2p

The carbon atom can use its two singly occupied p-type orbitals, to form two covalent bonds with two hydrogen atoms, yielding the singlet methylene CH_2, the simplest carbene. The carbon atom can also bond to four hydrogen atoms by an excitation of an electron from the doubly occupied 2s orbital to the empty 2p orbital, producing four singly occupied orbitals.

C*	↑↓	↑	↑	↑	↑
	1s	2s	2p	2p	2p

The energy released by formation of two additional bonds more than compensates for the excitation energy required, energetically favouring the formation of four C-H bonds.

Quantum mechanically, the lowest energy is obtained if the four bonds are equivalent, which requires that they are formed from equivalent orbitals on the carbon. A set of four equivalent orbitals
can be obtained that are linear combinations of the valence-shell (core orbitals are almost never
involved in bonding) s and p wave functions, which are the four sp³ hybrids.

C*	↑↓	↑	↑	↑	↑
	1s	sp³	sp³	sp³	sp³

In CH₄, four sp³ hybrid orbitals are overlapped by hydrogen 1s orbitals, yielding four σ (sigma)
bonds (that is, four single covalent bonds) of equal length and strength.

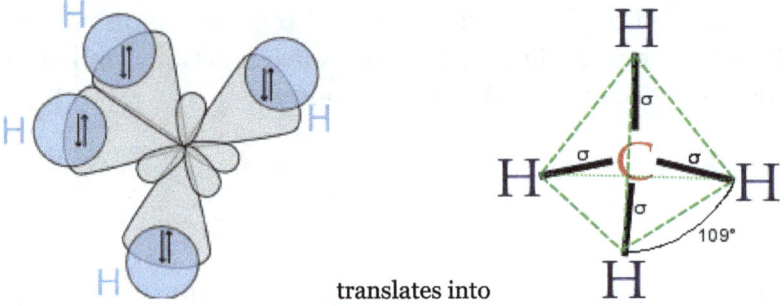

translates into

sp2

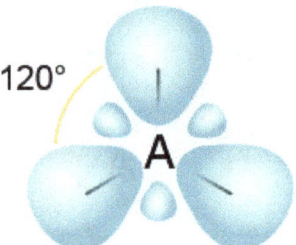

Three sp² orbitals.

Ethene structure

Other carbon based compounds and other molecules may be explained in a similar way. For example, ethene (C₂H₄) has a double bond between the carbons.

For this molecule, carbon sp² hybridises, because one π (pi) bond is required for the double bond
between the carbons and only three σ bonds are formed per carbon atom. In sp² hybridisation the
2s orbital is mixed with only two of the three available 2p orbitals,

C*	↑↓	↑	↑	↑	↑
	1s	sp²	sp²	sp²	2p

forming a total of three sp² orbitals with one remaining p orbital. In ethylene (ethene) the two carbon atoms form a σ bond by overlapping two sp² orbitals and each carbon atom forms two covalent bonds with hydrogen by s–sp² overlap all with 120° angles. The π bond between the carbon atoms perpendicular to the molecular plane is formed by 2p–2p overlap. The hydrogen–carbon bonds are all of equal strength and length, in agreement with experimental data.

sp

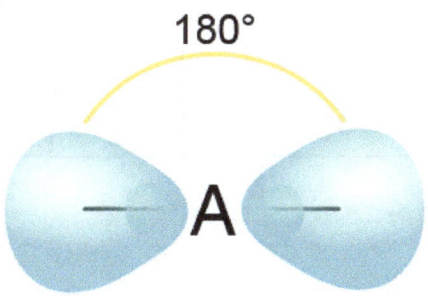

180°

Two sp orbitals

The chemical bonding in compounds such as alkynes with triple bonds is explained by sp hybridisation. In this model, the 2s orbital is mixed with only one of the three p orbitals,

C*	↑↓	↑	↑	↑	↑
	1s	sp	sp	2p	2p

resulting in two sp orbitals and two remaining p orbitals. The chemical bonding in acetylene (ethyne) (C_2H_2) consists of sp–sp overlap between the two carbon atoms forming a σ bond and two additional π bonds formed by p–p overlap. Each carbon also bonds to hydrogen in a σ s–sp overlap at 180° angles.

Hybridisation and Molecule Shape

Hybridisation helps to explain molecule shape, since the angles between bonds are (approximately) equal to the angles between hybrid orbitals, as explained above for the tetrahedral geometry of methane. As another example, the three sp² hybrid orbitals are at angles of 120° to each other, so this hybridisation favours trigonal planar molecular geometry with bond angles of 120°. Other examples are given in the table below.

Classification	Main group	Transition metal
AX_2	• Linear (180°) • sp hybridisation • E.g., CO_2	• Bent (90°) • sd hybridisation • E.g., VO_2^+
AX_3	• Trigonal planar (120°) • sp² hybridisation • E.g., BCl_3	• Trigonal pyramidal (90°) • sd² hybridisation • E.g., CrO_3

AX_4	• Tetrahedral (109.5°) • sp^3 hybridisation • E.g., CCl_4	• Tetrahedral (109.5°) • sd^3 hybridisation • E.g., MnO_4^-
AX_5		• Square pyramidal (73°, 123°) • sd^4 hybridisation • E.g., $Ta(CH_3)_5$
AX_6		• Trigonal prismatic (63.5°, 116.5°) • sd^5 hybridisation • E.g., $W(CH_3)_6$

Hybridisation of Hypervalent Molecules

Valence Shell Expansion

Hybridisation is often presented for main group AX_5 and above, as well as for many transition metal complexes, using the hybridisation scheme first proposed by Pauling.

Classification	Main group	Transition metal
AX_2		• Linear (180°) • sp hybridisation • E.g., $Ag(NH_3)_2^+$
AX_3		• Trigonal planar (120°) • sp^2 hybridisation • E.g., $Cu(CN)_3^{2-}$
AX_4		• Tetrahedral (109.5°) • sp^3 hybridisation • E.g., $Ni(CO)_4$ • Square planar (90°) • dsp^2 hybridisation • E.g., $PtCl_4^{2-}$
AX_5	• Trigonal bipyramidal (90°, 120°) • sp^3d hybridisation • E.g., PCl_5	Trigonal bipyramidal or Square pyramidal
AX_6	• Octahedral (90°) • sp^3d^2 hybridisation • E.g., SF_6	• Octahedral (90°) • d^2sp^3 hybridisation • E.g., $Mo(CO)_6$
AX_7	• Pentagonal bipyramidal (90°, 72°) • sp^3d^3 hybridisation • E.g., IF_7	Pentagonal bipyramidal, Capped octahedral or Capped trigonal prismatic

In this notation, d orbitals of main group atoms are listed after the s and p orbitals since they have the same principal quantum number (n), while d orbitals of transition metals are listed first since the s and p orbitals have a higher n. Thus for AX_6 molecules, sp^3d^2 hybridisation in the S atom involves 3s, 3p and 3d orbitals, while d^2sp^3 for Mo involves 4d, 5s and 5p orbitals.

Contrary Evidence

In 1990, Magnusson published a seminal work definitively excluding the role of d-orbital hybridisation in bonding in hypervalent compounds of second-row (period 3) elements, ending a point of contention and confusion. Part of the confusion originates from the fact that d-functions are essential in the basis sets used to describe these compounds (or else unreasonably high energies and distorted geometries result). Also, the contribution of the d-function to the molecular wavefunction is large. These facts were incorrectly interpreted to mean that d-orbitals must be involved in bonding.

For transition metal centres, the d and s orbitals are the primary valence orbitals, which are only weakly supplemented by the p orbitals. The question of whether the p orbitals actually participate in bonding has not been definitively resolved, but all studies indicate they play a minor role.

Resonance

In light of computational chemistry, a better treatment would be to invoke sigmaresonance in addition to hybridisation, which implies that each resonance structure has its own hybridisation scheme. For main group compounds, all resonance structures must obey the octet (8-electron) rule. For transition metal compounds, the resonance structures that obey the duodectet (12-electron) rule suffice to describe bonding, with optional inclusion of d^msp^n resonance structures.

Classification	Main group	Transition metal	
AX$_2$		Linear (180°)	
		(x2)	
AX$_3$		Trigonal planar (120°)	
		(x3)	

AX$_4$		**Tetrahedral (109.5°)** (x4) **Square planar (90°)** (x4) (x2)
AX$_5$	**Trigonal bipyramidal (90°, 120°)**	**Trigonal bipyramidal or Square pyramidal**
AX$_6$	**Octahedral (90°)**	**Octahedral (90°)**
AX$_7$	**Pentagonal bipyramidal (90°, 72°)** x30 other resonance structures	**Pentagonal bipyramidal, Capped octahedral or Capped trigonal prismatic**

Isovalent Hybridisation

Although ideal hybrid orbitals can be useful, in reality most bonds require orbitals of intermediate character. This requires an extension to include flexible weightings of atomic orbitals of each type (s, p, d) and allows for a quantitative depiction of bond formation when the molecular geometry deviates from ideal bond angles. The amount of p-character is not restricted to integer values; i.e., hybridisations like sp$^{2.5}$ are also readily described.

The hybridisation of bond orbitals is determined by Bent's rule: "Atomic s character concentrates in orbitals directed towards electropositive substituents".

Molecules with Lone Pairs

For molecules with lone pairs, the bonding orbitals are isovalent hybrids. For example, the two bond-forming hybrid orbitals of oxygen in water can be described as sp^4 to give the interorbital angle of 104.5°. This means that they have 20% s character and 80% p character and does *not* imply that a hybrid orbital is formed from one s and four p orbitals on oxygen since the 2p subshell of oxygen only contains three p orbitals. The shapes of molecules with lone pairs are:

- Trigonal pyramidal

    o Three isovalent hybrid bond orbitals

    o E.g., NH_3

- Bent

    o Two isovalent hybrid bond orbitals

    o . E.g., SO_2, H_2O

In such cases, there are two mathematically equivalent ways of representing lone pairs. They can be represented by orbitals of sigma and pi symmetry similar to molecular orbital theory or by equivalent orbitals similar to VSEPR theory.

Hypervalent Molecules

For hypervalent molecules with lone pairs, the bonding scheme can be split into a hypervalent component and a component consisting of isovalent bonding hybrids. The hypervalent component consists of resonating bonds using p orbitals. The table below shows how each shape is related to the two components and their respective descriptions.

Two		Number of isovalent bonding hybrids (marked in red)		
		One	—	
		Seesaw (AX_4E_1) (90°, 180°, >90°)	T-shaped (AX_3E_2) (90°, 180°)	Linear (AX_2E_3) (180°)
Hypervalent component	Linear axis (one p orbital)	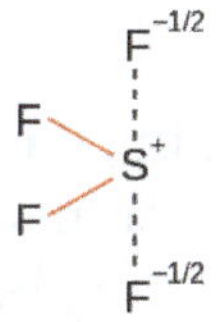	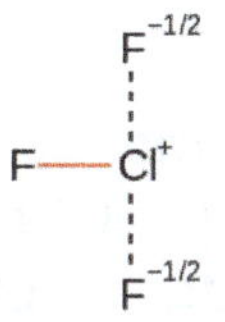	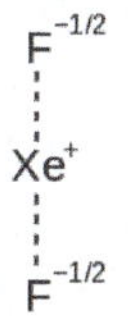
	Square planar equator (two p orbitals)		Square pyramidal (AX_5E_1) (90°, 90°) 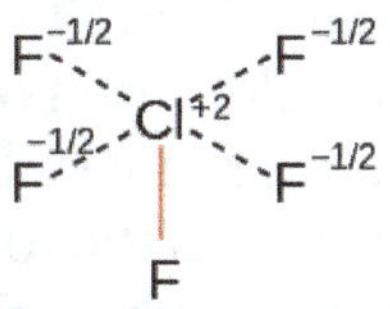	Square planar (AX_4E_2) (90°) 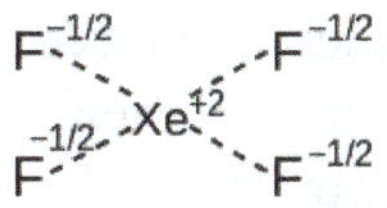
	Pentagonal planar equator (two p orbitals)		Pentagonal pyramidal (AX_6E_1) (90°, 72°) 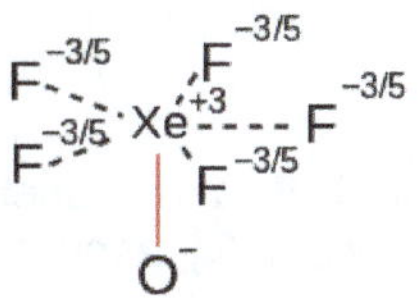	Pentagonal planar (AX_5E_2) (72°) 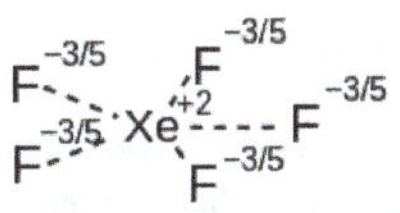

Hybridisation Defects

Hybridisation of s and p orbitals to form effective sp^x hybrids requires that they have comparable radial extent. While 2p orbitals are on average less than 10% larger than 2s, in part attributable to the lack of a radial node in 2p orbitals, 3p orbitals which have one radial node, exceed the 3s orbitals by 20–33%. The difference in extent of s and p orbitals increases further down a group. The hybridisation of atoms in chemical bonds can be analysed by considering localised molecular orbitals, for example using natural localised molecular orbitals in a natural bond orbital (NBO) scheme. In methane, CH_4, the calculated p/s ratio is approximately 3 consistent with "ideal" sp^3 hybridisation, whereas for silane, SiH_4, the p/s ratio is closer to 2. A similar trend is seen for the other 2p elements. Substitution of fluorine for hydrogen further decreases the p/s ratio. The 2p elements exhibit near ideal hybridisation with orthogonal hybrid orbitals. For heavier p block elements this assumption of orthogonality cannot be justified. These deviations from the ideal hybridisation were termed hybridisation defects by Kutzelnigg.

Photoelectron Spectra

One misconception concerning orbital hybridisation is that it incorrectly predicts the ultraviolet photoelectron spectra of many molecules. While this is true if Koopmans' theorem is applied to localised hybrids, quantum mechanics requires that the (in this case ionised) wavefunction obey the symmetry of the molecule which implies resonance in valence bond theory. For example, in methane, the ionised states (CH_4^+) can be constructed out of four resonance structures attributing the ejected electron to each of the four sp^3 orbitals. A linear combination of these four structures, conserving the number of structures, leads to a triply degenerate T_2 state and a A_1 state. The difference in energy between each ionised state and the ground state would be an ionisation energy, which yields two values in agreement with experiment.

Hybridisation Theory vs. Molecular Orbital Theory

Hybridisation theory is an integral part of organic chemistry and in general discussed together with molecular orbital theory. For drawing reaction mechanisms sometimes a classical bonding picture is needed with two atoms sharing two electrons. Predicting bond angles in methane with MO theory is not straightforward. Hybridisation theory explains bonding in alkenes and methane.

Bonding orbitals formed from hybrid atomic orbitals may be considered as localised molecular orbitals, which can be formed from the delocalised orbitals of molecular orbital theory by an appropriate mathematical transformation. For molecules with a closed electron shell in the ground state, this transformation of the orbitals leaves the total many-electron wave function unchanged. The hybrid orbital description of the ground state is therefore *equivalent* to the delocalised orbital description for ground state total energy and electron density, as well as the molecular geometry that corresponds to the minimum total energy value.

References

- Aue, W. P. and Bartholdi, E. and Ernst, R. R., Two-dimensional spectroscopy. Application to nuclear magnetic resonance; The Journal of Chemical Physics, 64, 2229-2246 (1976)

- Paula, Peter Atkins, Julio de (2009). Elements of physical chemistry (5th ed.). Oxford: Oxford U.P. p. 459. ISBN 978-0-19-922672-6

- "Background and Theory Page of Nuclear Magnetic Resonance Facility". Mark Wainwright Analytical Centre - University of Southern Wales Sydney. 9 December 2011. Retrieved 9 February 2014

- Greenwood, N. N.; & Earnshaw, A. (1997). Chemistry of the Elements (2nd Edn.), Oxford:Butterworth-Heinemann. ISBN 0-7506-3365-4

- Derrick, M.R., Stulik, D. and Landry J.M., Infrared Spectroscopy in Conservation Science, Scientific Tools for Conservation, Getty Publications, 2000. Retrieved December 11, 2015

- Gillespie, R.J. (2004), "Teaching molecular geometry with the VSEPR model", Journal of Chemical Education, 81 (3): 298–304, Bibcode:2004JChEd..81..298G, doi:10.1021/ed081p298

- Christian Laurence and Jean-François Gal "Lewis Basicity and Affinity Scales : Data and Measurement" Wiley, 2009. ISBN 978-0-470-74957-9

- James Keeler. "Chapter 2: NMR and energy levels" (reprinted at University of Cambridge). Understanding NMR Spectroscopy. University of California, Irvine. Retrieved 2007-05-11

- McMurray, J. (1995). Chemistry Annotated Instructors Edition (4th ed.). Prentice Hall. p. 272. ISBN 978-0-131-40221-8

- C. R. Landis; F. Weinhold (2007). "Valence and extra-valence orbitals in main group and transition metal bonding". Journal of Computational Chemistry. 28 (1): 198–203. doi:10.1002/jcc.20492

- "What Happened When We Took the SCiO Food Analyzer Grocery Shopping". IEEE Spectrum: Technology, Engineering, and Science News. Retrieved 2017-03-23

2

Synthesis of Organometallic Compounds

Elements in the periodic table are categorized into s, p, d and f blocks depending on the type of atomic orbital that have. Most organometallic compounds can be prepared by methods of synthesis of metals reacting with organic halide or through the process of hydrometallation, metathesis, and metal displacement. The major aspects of the elements in organometallic chemistry are discussed in this section.

Block (Periodic Table)

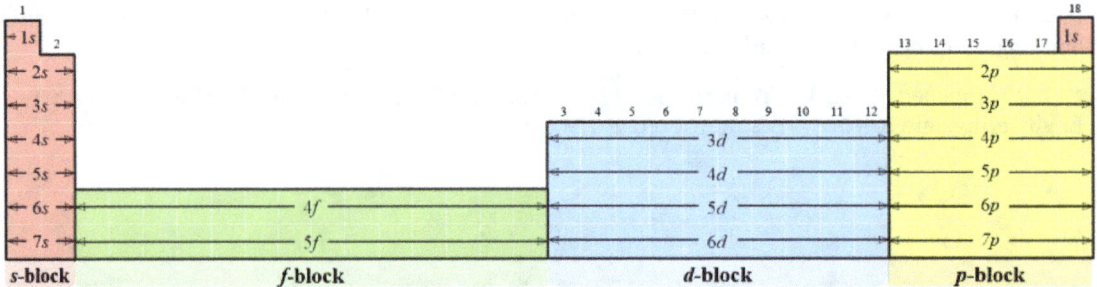

A block of the periodic table of elements is a set of adjacent groups. The term appears to have been first used by Charles Janet. The respective highest-energy electrons in each element in a block belong to the same atomic orbital type. Each block is named after its characteristic orbital; thus, the blocks are:

- s-block

- p-block

- d-block

- f-block

- g-block (hypothetical)

The block names (s, p, d, f and g) are derived from the spectroscopic notation for the associated atomic orbitals: sharp, principal, diffuse and fundamental, and then g which follows f in the alphabet.

The following is the order for filling the "subshell" orbitals, according to the Aufbau principle, which also gives the linear order of the "blocks" (as atomic number increases) in the periodic table:

1s, 2s, 2p, 3s, 3p, 4s, 3d, 4p, 5s, 4d, 5p, 6s, 4f, 5d, 6p, 7s, 5f, 6d, 7p, ...

For discussion of the nature of why the energies of the blocks naturally appear in this order in complex atoms, see atomic orbital and electron configuration.

The "periodic" nature of the filling of orbitals, as well as emergence of the s, p, d and f "blocks" is more obvious, if this order of filling is given in matrix form, with increasing principal quantum numbers starting the new rows ("periods") in the matrix. Then, each subshell (composed of the first two quantum numbers) is repeated as many times as required for each pair of electrons it may contain. The result is a compressed periodic table, with each entry representing two successive elements:

1s															
2s													2p	2p	2p
3s													3p	3p	3p
4s							3d	3d	3d	3d	3d	4p	4p	4p
5s							4d	4d	4d	4d	4d	5p	5p	5p
6s	4f	4f	4f	4f	4f	4f	4f	5d	5d	5d	5d	5d	6p	6p	6p
7s	5f	5f	5f	5f	5f	5f	5f	6d	6d	6d	6d	6d	7p	7p	7p

Periodic table

There is an approximate correspondence between this nomenclature of blocks, based on electronic configuration, and groupings of elements based on chemical properties. The s-block and p-block together are usually considered as the main group elements, the d-block corresponds to the transition metals, and the f-block are the lanthanides and the actinides. However, not everyone agrees on the exact membership of each set of elements, so that for example the Group 12 elements Zn, Cd and Hg are considered as main group by some scientists and transition metals by others. Groups (columns) in the f-block (between groups 2 and 3) are not numbered.

Helium is coloured differently from the p-block elements surrounding it because is in the s-block, with its outer (and only) electrons in the 1s atomic orbital, although its chemical properties are more similar to the p-block noble gases due to its full shell. In addition to the blocks listed in this table, there is a hypothetical g-block which is not pictured here. The g-block elements can be seen in the expanded extended periodic table. Also, lanthanum and actinium are placed under scandium and yttrium to reflect their status as d-block elements, as they have no electrons in the 4f and 5f orbitals, respectively, while lutetium and lawrencium do.

Blocks in the periodic table

Group →	1	2	3	4	5	6	7	8	9	10	11	12	13	14	15	16	17	18
↓ Period																		
1	1 H																	2 He
2	3 Li	4 Be											5 B	6 C	7 N	8 O	9 F	10 Ne

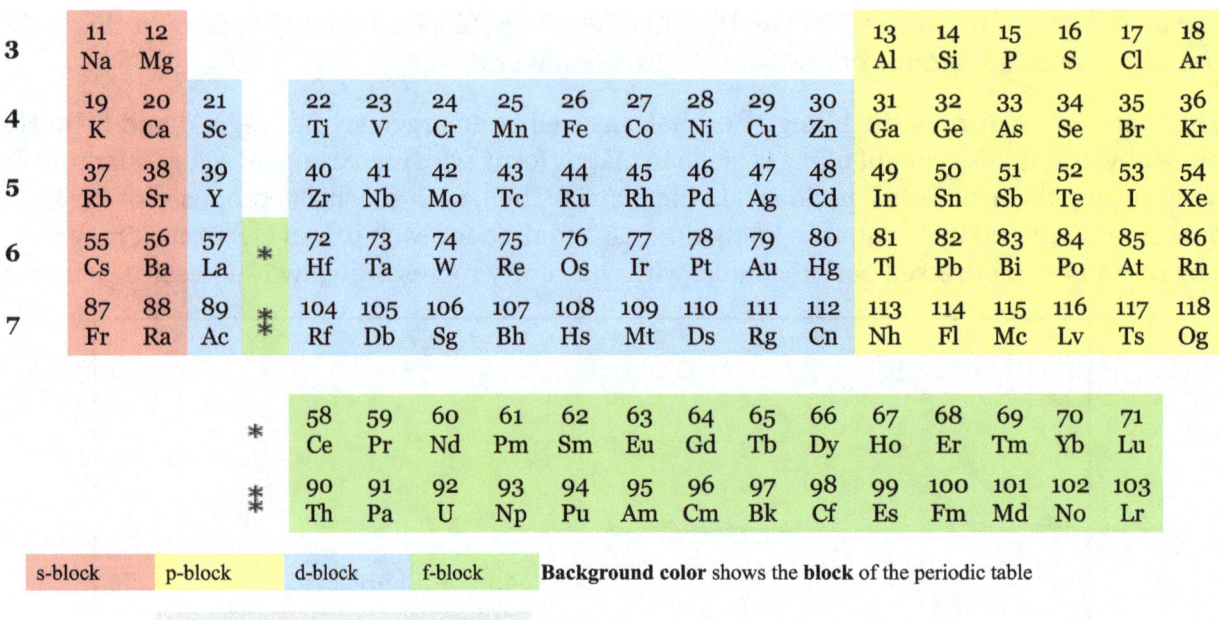

s-block   p-block   d-block   f-block   **Background color** shows the **block** of the periodic table

Primordial From decay   Synthetic   **Border** shows natural occurrence of the element

s-block

The s-block is on the left side of the periodic table that includes elements from the first two columns, the alkali metals (group 1) and alkaline earth metals (group 2), plus helium. Helium is a controversial element for the scientists as it can be placed in s block as well as p block too but most of the scientists consider it to be rest at the top of group 18 i.e. above neon(atomic number 10) as it has many properties similar to the group 18 elements.

Most s-block elements are highly reactive metals due to the ease with which their outer s-orbital electrons interact to form compounds. The first period elements in this block, however, are non-metals. Hydrogen is highly chemically reactive, like the other s-block elements, but helium is a virtually unreactive noble gas.

S-block elements are unified by the fact that their valence electrons (outermost electrons) are in the s orbital. The s-orbital is a single spherical cloud which can contain only one pair of electrons; hence, the s-block consists of only two columns in the periodic table. Elements in column 1, with a single s-orbital valence electron, are the most reactive of the block. Elements in the second column have two s-orbital valence electrons, and, except for helium, are only slightly less reactive.

p-block

The p-block is on the right side of the periodic table and includes elements from the six columns beginning with column 13 and ending with column 18. Helium, though being in the top of group 18, is not included in the p-block.

The p-block is home to the biggest variety of elements and is the only block that contains all three types of elements: metals, nonmetals, and metalloids. Generally, the p-block elements are best described in terms of element type or group.

P-block elements are unified by the fact that their valence electrons (outermost electrons) are in the p orbital. The p orbital consists of six lobed shapes coming off a central point at evenly spaced angles. The p orbital can hold a maximum of six electrons, hence there are six columns in the p-block. Elements in column 13, the first column of the p-block, have one p-orbital electron. Elements in column 14, the second column of the p-block, have two p-orbital electrons. The trend continues this way until we reach column 18, which has six p-orbital electrons.

Metals

P-block metals have classic metal characteristics: they are shiny, they are good conductors of heat and electricity, and they lose electrons easily. Generally, these metals have high melting points and readily react with nonmetals to form ionic compounds. Ionic compounds form when a positive metal ion bonds with a negative nonmetal ion.

Of the p-block metals, several have fascinating properties. Gallium, in the 3rd row of column 13, is a metal that can melt in the palm of a hand. Tin, in the fourth row of column 14, is an abundant, flexible, and extremely useful metal. It is an important component of many metal alloys like bronze, solder, and pewter.

Sitting right beneath tin is lead, a toxic metal. Ancient people used lead for a variety of things, from food sweeteners to pottery glazes to eating utensils. It has been suspected that lead poisoning is related to the fall of Roman civilization, but further research has shown this to be unlikely. For a long time, lead was used in the manufacturing of paints. It was only within the last century that lead paint use has been restricted due to its toxic nature.

Metalloids

Metalloids have properties of both metals and nonmetals, but the term 'metalloid' lacks a strict definition. All of the elements that are commonly recognized as metalloids are in the p-block: boron, silicon, germanium, arsenic, antimony, and tellurium. Metalloids tend to have lower electrical conductivity than metals, yet often higher than nonmetals. They tend to form chemical bonds similarly to nonmetals, but may dissolve in metallic alloys without covalent or ionic bonding. Metalloid additives can improve properties of metallic alloys, sometimes paradoxically to their own apparent properties. Some may give a better electrical conductivity, higher corrosion resistance, ductility, or fluidity in molten state, etc. to the alloy.

Boron has many carbon-like properties, but is very rare. It has many uses, for example a P type semiconductor dopant.

Silicon is perhaps the most famous metalloid. It is the second most abundant element in Earth's crust and one of the main ingredients in glass. It is used to make microchips for computers and other electronic devices. It is also used in certain metallic alloys, e.g. to improve casting properties of alumimium. So valuable is silicon to the technology industry that Silicon Valley in California is named after it.

Germanium has properties very similar to silicon, yet this element is much more rare. It was once used for its semiconductor properties pretty much as silicon is now, and it has some superior properties at that, but is now a rare material in the industry.

Arsenic is a toxic metalloid that has been used throughout history as an additive to metal alloys, paints, and even makeup.

Antimony is used as a constituent in casting alloys such as printing metal.

- Not always considered as metalloids:

    o   Carbon, in the same column with silicon and germanium, electrically fairly conductive unlike most other nonmetals, and to an extent preferred as a trace constituent in certain metallic alloys such as steel.

    o   Phosphorus has metallurgical uses among others, e.g. a constituent of some copper alloys.

    o   Selenium, once used as a semiconductor material, and also used to improve properties of metallic alloys.

    o   Aluminium is generally considered a metal, but it has some metalloid/non-metal properties such as negative oxidation states.

Noble Gases

Previously called inert gases, their name was changed as there are a few other gases that are inert but not noble gases, such as nitrogen. The noble gases are located in the far right column of the periodic table, also known as Group Zero or Group Eighteen. Noble gases are also called as aerogens but this nomenclature of the group is not officially accepted by the IUPAC.

All of the noble gases have full outer shells with eight electrons. However, at the top of the noble gases is helium, with a shell that is full with only two electrons. The fact that their outer shells are full means they rarely react with other elements, which led to their original title of "inert."

Because of their chemical properties, these gases are also used in the laboratory to help stabilize reactions that would usually proceed too quickly. As the atomic numbers increase, the elements become rarer. They are not just rare in nature, but rare as useful elements, too.

- Helium is best known for its low density, used to safely produce buoyancy for zeppelins and balloons.

- Neon is notorious as the red to yellow glow medium of old low power signal lamps and signs.

- Argon is used as a protective gas in MIG and TIG welding.

- Xenon is used as a plasma medium in high intensity arc lamps with tungsten electrodes. Automotive xenon lights, however, are mostly mercury vapor bulbs with low pressure xenon to help striking the arc and producing light instantly.

- Krypton has many uses like arc flash medium. Krypton filled incandescent bulbs were once the most efficient variety, before being replaced by halogen technology.

- Radon is radioactive, and one of the densest elements to remain in gas state at room temperature.

Halogens

The second column from the right side of the periodic table, group 17, is the halogen family of elements. These elements are all just one electron shy of having full shells. Because they are so close to being full, they have the trait of combining with many different elements and are very reactive. They are often found bonding with metals and elements from Group One, as these elements in each have one electron.

Not all halogens react with the same intensity. Fluorine is the most reactive and combines with most elements from around the periodic table. As with other columns, reactivity decreases as the atomic number increases.

When a halogen combines with another element, the resulting compound is called a halide. One of the best examples of a halide is sodium chloride (NaCl).

d-block

The d-block is on the middle of the periodic table and includes elements from columns 3 through 12. These elements are also known as the transition metals because they show a transitivity in their properties i.e. they show a trend in their properties.

The d-block elements are all metals which exhibit two or more ways of forming chemical bond. Because there is a relatively small difference in the energy of the different d-orbital electrons, the number of electrons participating in chemical bonding can vary. This results in the same element exhibiting two or more oxidation states, which determines the type and number of its nearest neighbors in chemical compounds.

D-block elements are unified by having in their outermost electrons one or more d-orbital electrons but no p-orbital electrons. The d-orbitals can contain up to five pairs of electrons; hence, the block includes ten columns in the periodic table.

f-block

The f-block is in the center-left of a 32-column periodic table but in the footnoted appendage of 18-column tables. These elements are not generally considered as part of any group. They are often called inner transition metals because they provide a transition between the s-block and d-block in the 6th and 7th row (period), in the same way that the d-block transition metals provide a transitional bridge between the s-block and p-block in the 4th and 5th rows.

The known f-block elements come in two series, the lanthanides of period 6 and the radioactive actinides of period 7. All are metals. Because the f-orbital electrons are less active in determining the chemistry of these elements, their chemical properties are mostly determined by outer s-orbital electrons. Consequently, there is much less chemical variability within the f-block than within the s-, p-, or d-blocks.

F-block elements are unified by having one or more of their outermost electrons from in the f-orbital but none in the d-orbital or p-orbital. The f-orbitals can contain up to seven pairs of electrons; hence, the block includes fourteen columns in the periodic table.

General Methods of Preparation

An organometallic compound contains one or more metal-carbon bonds.

Synthesis

General Methods of Preparation

Most organometallic compounds can be synthesized by using one of four M-C bond forming reactions of a metal with an organic halide, metal displacement, metathesis and hydrometallation.

a)   Reaction with metal and transmetallation

The net reaction of an electropositive metal M and a halogen-substituted hydrocarbon is

$$2M + RX\,(alkyl\ or\ aryl) \rightarrow MR + MX$$
$$8Li + 4CH_3Cl \rightarrow Li_4(CH_3)_4 + 4LiCl$$
$$Mg + CH_3Br \rightarrow CH_3MgBr\,(organometal\ halide\ with\ Mg, Al, Zn)$$

If, one metal atom takes the place of another, it is called transmetallation

$$M + M'R \rightarrow M' + MR$$
$$2Ga + 3CH_{3-}Hg^-CH_3 \rightarrow 3Hg + 2Ga(CH_3)_3$$

Transmetallation is favorable when the displacing metal is higher in the electrochemical series than the displaced metal.

b)   Metathesis

The metathesis of an organometallic compound MR and a binary halide EX is a widely used synthetic route in organometallic chemistry.

$$MR + EX \rightarrow ER + MX$$
$$Li_4(CH_3)_4 + SiCl \rightarrow 4LiCl + Si(CH_3)_4$$
$$Al_2(CH_3)_6 + 2BF_3 \rightarrow 2AlF_3 + 2B(CH_3)_3$$

Metathesis reaction can frequently be predicted from electronegativity or hard and soft acid-base considerations.

Hydrocarbon groups tends to bond to the more electronegative element; the halogen favors the formation of ionic compounds with the more electropositive metal.

In brief, the alkyl and aryl group tends to migrate from the less to the more electronegative element [χ = electronegativity].

$$MR + EX \rightarrow ER + MX$$

$$\begin{array}{cccccccc} Li & Mg & Al & Zn & Si & B & As & P \end{array}$$

$$x: 0.98 \;\; 1.31 \;\; 1.61 \;\; 1.66 \;\; 1.90 \;\; 2.04 \;\; 2.18 \;\; 2.19$$

When the electronegativities are similar, the correct outcome may be predicted, with care*, by considering the combination of the softer element with organic group and harder element with fluoride or chloride.

*An insoluble product or reactant may change the outcome, e.g.;

$$SnPh_4(THF) + HgBr_2(THF) \rightarrow HgPhBr(s) + Ph_3SnBr(THF)$$

HgPhBr turns out to be insoluble in THF

Metathesis reactions involving the same central element are often referred to as redistribution reactions.

$$SiCl_4 + SiMe_4 \rightarrow Me_3SiCl + Me_3SiCl + Me_2SiCl_2 + \ldots\ldots$$

$$3GeCl_4 + 2Al_2Me_6 \rightarrow 3GeMe_4 + 4AlCl_3$$

Al is more electropositive than Ge, this reaction occurs as it is thermodynamically favorable.

c)   Hydrometallation

The net outcome of the addition of a metal hydride to an alkene is an alkylmetal compound.

$$EH + H_2C = CH_2 \rightarrow E - CH_2-CH_3$$

The reaction is driven by the high strength of E-C bond relative to that of most E-H bonds, and occurs with a wide variety of compounds that contain E-H bonds.

Hydroboration

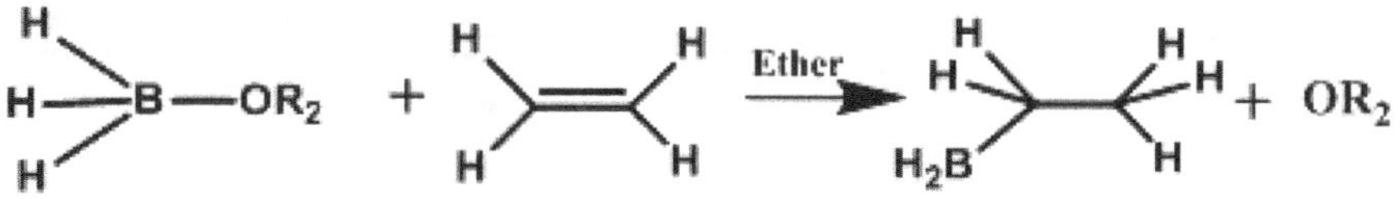

Hydrosilylation

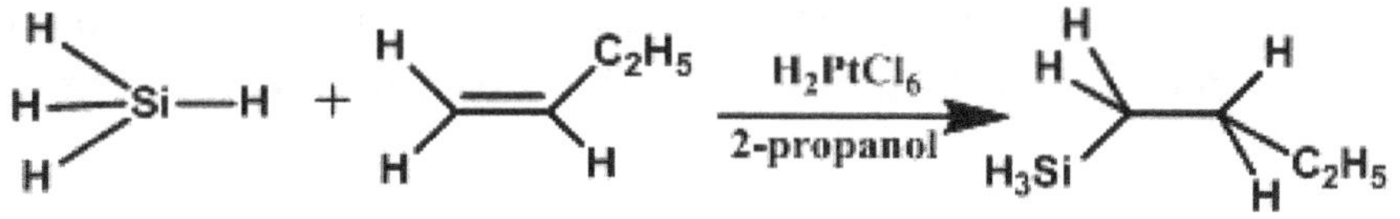

Ionic and electron-deficient compounds of Group 1, 2

Organometallic derivatives of all Group 1 metals are known. Amongst, the alkyllithium compounds are most thoroughly studied and useful reagents.

Many of them are commercially available.

MeLi is generally handled in ether solution, but RLi compounds with longer chains are soluble in hydrocarbons.

Commercial preparation:

$$M + RX \rightarrow MR \text{ (often contaminated with halide)}$$

The best method would be:

$$HgR_2 + 2Li \rightarrow 2LiR + Hg$$

MeLi exists as a tetrahedral cluster in the solid state and in the solution. Many of its higher homologs exist in solution as hexamers or equilibrium mixture of aggregates ranging up to haxamers.

The larger aggregates can be broken down by Lewis bases, such as, TMEDA.

Common organolithium compounds have one Li per organic group.

Several polylithiated organic molecules containing several lithium atoms per molecule are known.

The simplest example is Li_2CH_2, which can be prepared by the pyrolysis of MeLi which crystallizes in a distorted antifulorite* structure. However, the finer details of the orientation of the CH_2 groups are yet to be established.

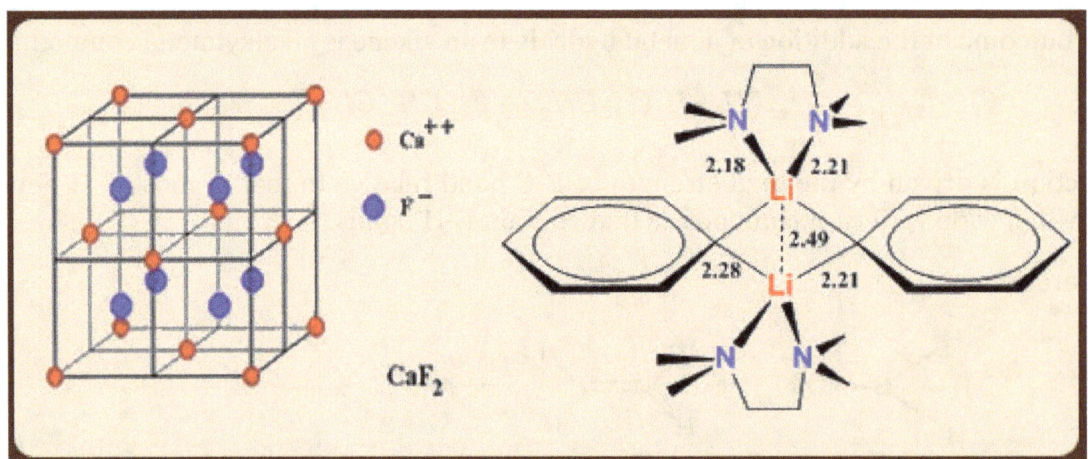

*the antifluorite structure is the inverse of the fluorite structure in which the locations ofcations and anions are reversed. Look into the structures of CaF_2 (fluorite structure) and K_2O (antifluorite structure). An fcc array of cations and all the tetrahedral holes are filled with anions.

Ion

An ion is an atom or a molecule in which the total number of electrons is not equal to the total number of protons, giving the atom or molecule a net positive or negative electrical charge. Ions can be created, by either chemical or physical means, via ionization.

In chemical terms, if a neutral atom loses one or more electrons, it has a net positive charge and is known as a cation.

If an atom gains electrons, it has a net negative charge and is known as an anion.

Ions consisting of only a single atom are atomic or monatomic ions; if they consist of two or more atoms, they are molecular or polyatomic ions. Because of their electric charges, cations and anions attract each other and readily form ionic compounds, such as salts.

In the case of physical ionization of a medium, such as a gas, which are known as "ion pairs" are created by ion impact, and each pair consists of a free electron and a positive ion.

History of Discovery

The word *ion* comes from the Greek word, *ion*, "going", the present participle of *ienai*, "to go". This term was introduced by English physicist and chemist Michael Faraday in 1834 for the then-unknown species that *goes* from one electrode to the other through an aqueous medium. Faraday did not know the nature of these species, but he knew that since metals dissolved into and entered a solution at one electrode, and new metal came forth from a solution at the other electrode, that some kind of substance moved through the solution in a current, conveying matter from one place to the other.

Faraday also introduced the words *anion* for a negatively charged ion, and *cation* for a positively charged one. In Faraday's nomenclature, cations were named because they were attracted to the cathode in a galvanic device and anions were named due to their attraction to the anode.

Svante Arrhenius put forth, in his 1884 dissertation, his explanation of the fact that solid crystalline salts disassociate into paired charged particles when dissolved, for which he would win the 1903 Nobel Prize in Chemistry. Arrhenius' explanation was that in forming a solution, the salt dissociates into Faraday's ions. Arrhenius proposed that ions formed even in the absence of an electric current.

Characteristics

Ions in their gas-like state are highly reactive, and do not occur in large amounts on Earth, except in flames, lightning, electrical sparks, and other plasmas.

These gas-like ions rapidly interact with ions of opposite charge to give neutral molecules or ionic salts. Ions are also produced in the liquid or solid state when salts interact with solvents (for example, water) to produce "solvated ions," which are more stable, for reasons involving a combination of energy and entropy changes as the ions move away from each other to interact with the liquid. These stabilized species are more commonly found in the environment at low temperatures. A common example is the ions present in seawater, which are derived from the dissolved salts.

All ions are charged, which means that like all charged objects they are:

- attracted to opposite electric charges (positive to negative, and vice versa),

- repelled by like charges,

- when moving, travel in trajectories that are deflected by a magnetic field.

Electrons, due to their smaller mass and thus larger space-filling properties as matter waves, determine the size of atoms and molecules that possess any electrons at all. Thus, anions (negatively charged ions) are larger than the parent molecule or atom, as the excess electron(s) repel each other, and add to the physical size of the ion, because its size is determined by its electron cloud. As such, in general, cations are smaller than the corresponding parent atom or molecule due to the smaller size of its electron cloud. One particular cation (that of hydrogen) contains no electrons, and thus consists of a single proton - *very much smaller* than the parent hydrogen atom.

Anions and Cations

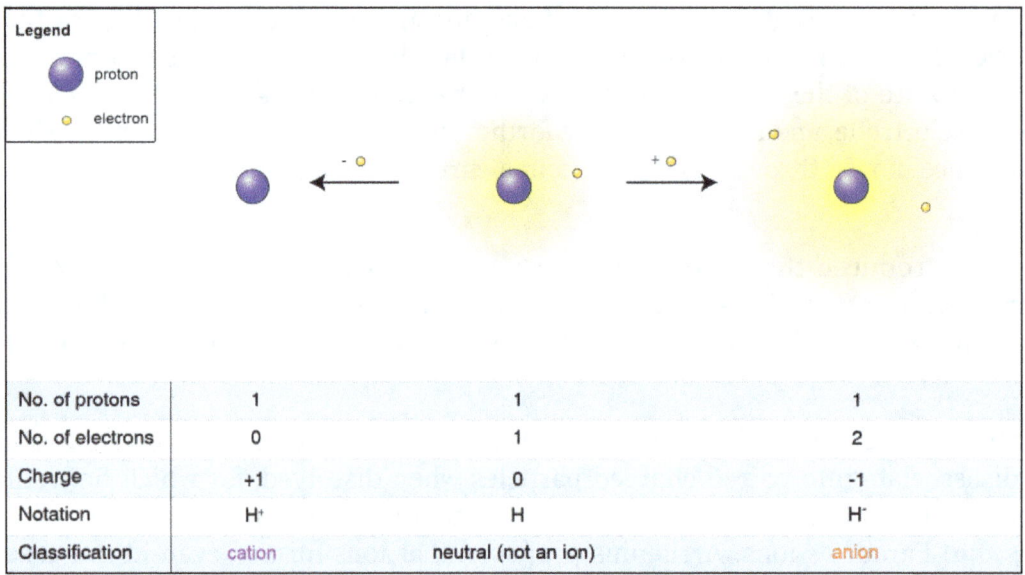

Legend		
● proton		
○ electron		

No. of protons	1	1	1
No. of electrons	0	1	2
Charge	+1	0	-1
Notation	H$^+$	H	H$^-$
Classification	cation	neutral (not an ion)	anion

Hydrogen atom (centre) contains a single proton and a single electron. Removal of the electron gives a cation (left), whereas addition of an electron gives an anion (right). The hydrogen anion, with its loosely held two-electron cloud, has a larger radius than the neutral atom, which in turn is much larger than the bare proton of the cation. Hydrogen forms the only cation that has no electrons, but even cations that (unlike hydrogen) still retain one or more electrons are still smaller than the neutral atoms or molecules from which they are derived.

Since the electric charge on a proton is equal in magnitude to the charge on an electron, the net electric charge on an ion is equal to the number of protons in the ion minus the number of electrons.

An anion (−), from the Greek word (*ánō*), meaning "up", is an ion with more electrons than protons, giving it a net negative charge (since electrons are negatively charged and protons are positively charged).

A cation (+), from the Greek word (*katá*), meaning "down", is an ion with fewer electrons than protons, giving it a positive charge.

There are additional names used for ions with multiple charges. For example, an ion with a

−2 charge is known as a dianion and an ion with a +2 charge is known as a dication. A zwitterion is a neutral molecule with positive and negative charges at different locations within that molecule.

Cations and ions are measured by their ionic radius and they differ in relative size: "Cations are small, most of them less than 10^{-8} cm in radius. But most anions are large, as is the most common Earth anion, oxygen. From this fact it is apparent that most of the space of a crystal is occupied by the anion and that the cations fit into the spaces between them."

In terms of an angstrom Å, a cation has radius less than 8 Å while an anion has radius greater than 1.3 Å.

Natural Occurrences

Ions are ubiquitous in nature and are responsible for diverse phenomena from the luminescence of the Sun to the existence of the Earth's ionosphere. Atoms in their ionic state may have a different colour from neutral atoms, and thus light absorption by metal ions gives the colour of gemstones. In both inorganic and organic chemistry (including biochemistry), the interaction of water and ions is extremely important; an example is the energy that drives breakdown of adenosine triphosphate (ATP). The following sections describe contexts in which ions feature prominently; these are arranged in decreasing physical length-scale, from the astronomical to the microscopic.

Astronomical

A collection of non-aqueous gas-like ions, or even a gas containing a proportion of charged particles, is called a plasma. Greater than 99.9% of visible matter in the Universe may be in the form of plasmas. These include our Sun and other stars and the space between planets, as well as the space in between stars. Plasmas are often called the *fourth state of matter* because their properties are substantially different from those of solids, liquids, and gases. Astrophysical plasmas predominantly contain a mixture of electrons and protons (ionized hydrogen).

Related Technology

Ions can be non-chemically prepared using various ion sources, usually involving high voltage or temperature. These are used in a multitude of devices such as mass spectrometers, optical emission spectrometers, particle accelerators, ion implanters, and ion engines.

As reactive charged particles, they are also used in air purification by disrupting microbes, and in household items such as smoke detectors.

As signalling and metabolism in organisms are controlled by a precise ionic gradient across membranes, the disruption of this gradient contributes to cell death. This is a common mechanism exploited by natural and artificial biocides, including the ion channelsgramicidin and amphotericin (a fungicide).

Inorganic dissolved ions are a component of total dissolved solids, an indicator of water quality in the world.

Detection of Ionizing Radiation

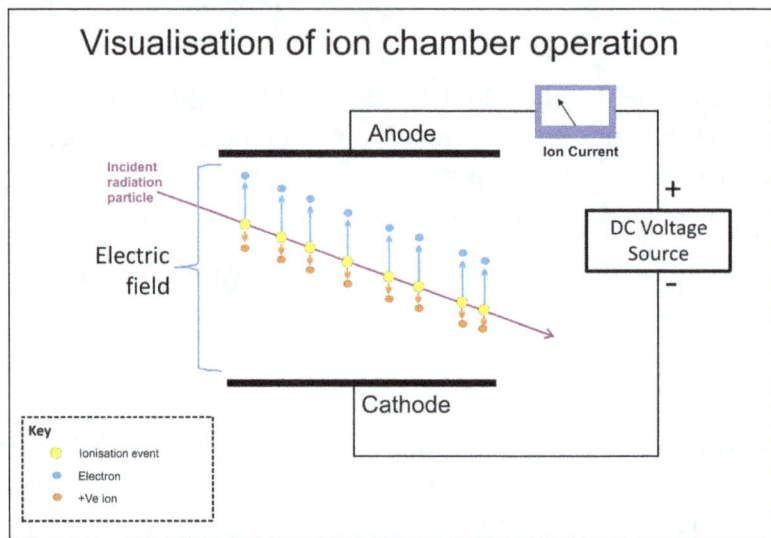

Schematic of an ion chamber, showing drift of ions. Electrons drift faster than positive ions due to their much smaller mass.

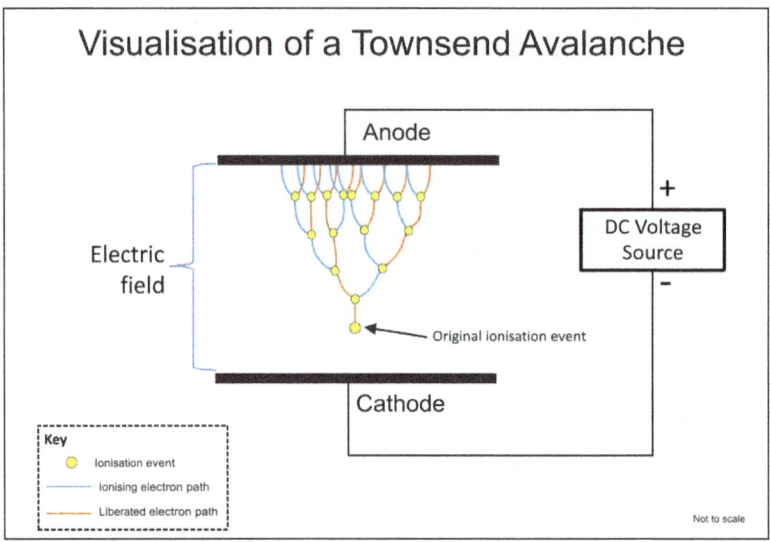

Avalanche effect between two electrodes. The original ionization event liberates one electron, and each subsequent collision liberates a further electron, so two electrons emerge from each collision: the ionizing electron and the liberated electron.

The ionizing effect of radiation on a gas is extensively used for the detection of radiation such as alpha, beta, gamma and X-rays. The original ionization event in these instruments results in the formation of an "ion pair"; a positive ion and a free electron, by ion impact by the radiation on the gas molecules. The ionization chamber is the simplest of these detectors, and collects all the charges created by *direct ionization* within the gas through the application of an electric field.

The Geiger–Müller tube and the proportional counter both use a phenomenon known as a Townsend avalanche to multiply the effect of the original ionizing event by means of a cascade effect whereby the free electrons are given sufficient energy by the electric field to release further electrons by ion impact.

Chemistry

Denoting the Charged State

$$Fe^{2+} \quad Fe^{++} \quad Fe^{\oplus\oplus}$$

Equivalent notations for an iron atom (Fe) that lost two electrons, referred to as ferrous.

When writing the chemical formula for an ion, its net charge is written in superscript immediately after the chemical structure for the molecule/atom. The net charge is written with the magnitude *before* the sign; that is, a doubly charged cation is indicated as 2+ instead of +2. However, the magnitude of the charge is omitted for singly charged molecules/atoms; for example, the sodium cation is indicated as Na^+ and *not* Na^{1+}.

An alternative (and acceptable) way of showing a molecule/atom with multiple charges is by drawing out the signs multiple times; this is often seen with transition metals. Chemists sometimes circle the sign; this is merely ornamental and does not alter the chemical meaning. All three representations of $Fe2^+$ shown in the figure are, thus, equivalent.

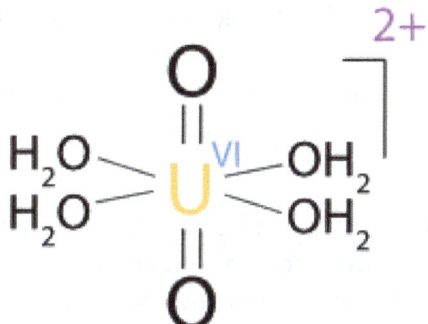

Mixed Roman numerals and charge notations for the uranyl ion. The oxidation state of the metal is shown as superscripted Roman numerals, whereas the charge of the entire complex is shown by the angle symbol together with the magnitude and sign of the net charge.

Monatomic ions are sometimes also denoted with Roman numerals; for example, the Fe2+ example seen above is occasionally referred to as Fe(II) or Fe^{II}. The Roman numeral designates the *formal oxidation state* of an element, whereas the superscripted numerals denote the net charge. The two notations are, therefore, exchangeable for monatomic ions, but the Roman numerals *cannot* be applied to polyatomic ions. However, it is possible to mix the notations for the individual metal centre with a polyatomic complex, as shown by the uranyl ion example.

Sub-classes

If an ion contains unpaired electrons, it is called a *radical* ion. Just like uncharged radicals, radical ions are very reactive. Polyatomic ions containing oxygen, such as carbonate and sulfate, are called *oxyanions*. Molecular ions that contain at least one carbon to hydrogen bond are called *organic ions*. If the charge in an organic ion is formally centred on a carbon, it is termed a *carbocation* (if positively charged) or *carbanion* (if negatively charged).

Formation

Formation of Monatomic Ions

Monatomic ions are formed by the gain or loss of electrons to the valence shell (the outer-most electron shell) in an atom. The inner shells of an atom are filled with electrons that are tightly bound to the positively charged atomic nucleus, and so do not participate in this kind of chemical interaction. The process of gaining or losing electrons from a neutral atom or molecule is called *ionization.*

Atoms can be ionized by bombardment with radiation, but the more usual process of ionization encountered in chemistry is the transfer of electrons between atoms or molecules. This transfer is usually driven by the attaining of stable ("closed shell") electronic configurations. Atoms will gain or lose electrons depending on which action takes the least energy.

For example, a sodium atom, Na, has a single electron in its valence shell, surrounding 2 stable, filled inner shells of 2 and 8 electrons. Since these filled shells are very stable, a sodium atom tends to lose its extra electron and attain this stable configuration, becoming a sodium cation in the process:

$$Na \rightarrow Na^+ + e^-$$

On the other hand, a chlorine atom, Cl, has 7 electrons in its valence shell, which is one short of the stable, filled shell with 8 electrons. Thus, a chlorine atom tends to *gain* an extra electron and attain a stable 8-electron configuration, becoming a chloride anion in the process:

$$Cl + e^- \rightarrow Cl^-$$

This driving force is what causes sodium and chlorine to undergo a chemical reaction, wherein the "extra" electron is transferred from sodium to chlorine, forming sodium cations and chloride anions. Being oppositely charged, these cations and anions form ionic bonds and combine to form sodium chloride, NaCl, more commonly known as table salt.

$$Na^+ + Cl^- \rightarrow NaCl$$

Formation of Polyatomic and Molecular Ions

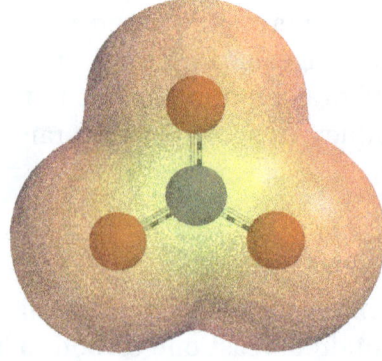

An electrostatic potential map of the nitrate ion (NO_3^-). The 3-dimensional shell represents a single arbitrary isopotential.

Polyatomic and molecular ions are often formed by the gaining or losing of elemental ions such as a proton, H^+, in neutral molecules. For example, when ammonia, NH_3, accepts a proton, H^+—a process called protonation—it forms the ammonium ion, NH_4^+. Ammonia and ammonium have the same number of electrons in essentially the same electronic configuration, but ammonium has an extra proton that gives it a net positive charge.

Ammonia can also lose an electron to gain a positive charge, forming the ion $\cdot NH_3^+$. However, this ion is unstable, because it has an incomplete valence shell around the nitrogen atom, making it a very reactive radical ion.

Due to the instability of radical ions, polyatomic and molecular ions are usually formed by gaining or losing elemental ions such as H^+, rather than gaining or losing electrons. This allows the molecule to preserve its stable electronic configuration while acquiring an electrical charge.

Ionization Potential

The energy required to detach an electron in its lowest energy state from an atom or molecule of a gas with less net electric charge is called the *ionization potential*, or *ionization energy*. The nth ionization energy of an atom is the energy required to detach its nth electron after the first $n - 1$ electrons have already been detached.

Each successive ionization energy is markedly greater than the last. Particularly great increases occur after any given block of atomic orbitals is exhausted of electrons. For this reason, ions tend to form in ways that leave them with full orbital blocks. For example, sodium has one *valence electron* in its outermost shell, so in ionized form it is commonly found with one lost electron, as Na^+. On the other side of the periodic table, chlorine has seven valence electrons, so in ionized form it is commonly found with one gained electron, as Cl^-. Caesium has the lowest measured ionization energy of all the elements and helium has the greatest. In general, the ionization energy of metals is much lower than the ionization energy of nonmetals, which is why, in general, metals will lose electrons to form positively charged ions and nonmetals will gain electrons to form negatively charged ions.

Ionic Bonding

Ionic bonding is a kind of chemical bonding that arises from the mutual attraction of oppositely charged ions. Ions of like charge repel each other, and ions of opposite charge attract each other. Therefore, ions do not usually exist on their own, but will bind with ions of opposite charge to form a crystal lattice. The resulting compound is called an *ionic compound*, and is said to be held together by *ionic bonding*. In ionic compounds there arise characteristic distances between ion neighbours from which the spatial extension and the ionic radius of individual ions may be derived.

The most common type of ionic bonding is seen in compounds of metals and nonmetals (except noble gases, which rarely form chemical compounds). Metals are characterized by having a small number of electrons in excess of a stable, closed-shell electronic configuration. As such, they have the tendency to lose these extra electrons in order to attain a stable configuration. This property is known as *electropositivity*. Non-metals, on the other hand, are characterized by having an electron configuration just a few electrons short of a stable configuration. As such, they have the tendency to gain more electrons in order to achieve a stable configuration. This tendency is known as *electronegativity*. When a highly electropositive metal is combined with a highly electronegative nonmetal, the extra electrons from the metal atoms are transferred to the electron-deficient nonmetal atoms. This reaction produces metal cations and nonmetal anions, which are attracted to each other to form a *salt*.

Common Ions

Common cations		
Common name	**Formula**	**Historic name**
Simple cations		
Aluminium	Al^{3+}	
Barium	Ba^{2+}	
Beryllium	Be^{2+}	
Calcium	Ca^{2+}	
Chromium(III)	Cr^{3+}	
Copper(I)	Cu^+	cuprous
Copper(II)	Cu^{2+}	cupric
Hydrogen	H^+	
Iron(II)	Fe^{2+}	ferrous
Iron(III)	Fe^{3+}	ferric
Lead(II)	Pb^{2+}	plumbous
Lead(IV)	Pb^{4+}	plumbic
Lithium	Li^+	
Magnesium	Mg^{2+}	
Manganese(II)	Mn^{2+}	
Mercury(II)	Hg^{2+}	mercuric
Potassium	K^+	kalic
Silver	Ag^+	argentous
Sodium	Na^+	natric
Strontium	Sr^{2+}	
Tin(II)	Sn^{2+}	stannous
Tin(IV)	Sn^{4+}	stannic
Zinc	Zn^{2+}	
Polyatomic cations		
Ammonium	NH_4^+	
Hydronium	H_3O^+	
Mercury(I)	Hg_2^{2+}	mercurous

Common anions		
Formal name	**Formula**	**Alt. name**
Simple anions		
Azide	N_3^-	
Bromide	Br^-	
Chloride	Cl^-	
Fluoride	F^-	
Hydride	H^-	
Iodide	I^-	
Nitride	N^{3-}	
Oxide	O^{2-}	
Sulfide	S^{2-}	
Oxoanions		
Carbonate	CO_3^{2-}	
Chlorate	ClO_3^-	
Chromate	CrO_4^{2-}	
Dichromate	$Cr_2O_7^{2-}$	
Dihydrogen phosphate	$H_2PO_4^-$	
Hydrogen carbonate	HCO_3^-	bicarbonate
Hydrogen sulfate	HSO_4^-	bisulfate
Hydrogen sulfite	HSO_3^-	bisulfite
Hydroxide	OH^-	
Hypochlorite	ClO^-	
Monohydrogen phosphate	HPO_4^{2-}	
Nitrate	NO_3^-	
Nitrite	NO_2^-	
Perchlorate	ClO_4^-	
Permanganate	MnO_4^-	
Peroxide	O_2^{2-}	
Phosphate	PO_4^{3-}	
Sulfate	SO_4^{2-}	
Sulfite	SO_3^{2-}	
Superoxide	O_2^-	
Thiosulfate	$S_2O_3^{2-}$	
Silicate	SiO_4^{4-}	
Metasilicate	SiO_3^{2-}	
Aluminium silicate	$AlSiO_4^-$	
Anions from organic acids		
Acetate	CH_3COO^-	ethanoate
Formate	$HCOO^-$	methanoate
Oxalate	$C_2O_4^{2-}$	ethanedioate
Cyanide	CN^-	

Radical Anion Salts

Sodium naphthalide is an example of an organometallic salt with a delocalized radical anion, $C_{10}H_8^-$.

Such compounds are readily prepared by reacting an aromatic compound with an alkali metal in a polar aprotic solvent.

Naphthalene dissolved in THF reacts with Na metal to produce a dark green solution of sodium naphthalide.

$$Na(s) + C_{10}H_8 (THF) \longrightarrow Na[C_{10}H_8](THF)$$

EPR spectra show that the odd electron is delocalized in an antibonding orbital of $C_{10}H_8$.

Formation of radical anion is more favorable when the π of LUMO of the arene is low in energy.

Simple MOT predicts that the energy of LUMO decreases steadily on going from benzene to more extensively conjugated hydrocarbons.

Sodium naphthalide and similar compounds are highly reactive reducing agents.

They are preferred to sodium because unlike sodium, they are readily soluble in ethers.

The resulting homogeneous reaction is generally faster and easier to control than a heterogeneous reaction between one reagent in solution and pieces of sodium metal, which are often coated with unreactive sodium oxide or with insoluble reaction products.

The additional advantage is that by proper choice of the aromatic group the reduction potential of the reagent can be chosen to match the requirements of a particular synthetic task.

Alternative route to delocalized anion is the reductive cleavage of acidic C—H bonds by an alkali metal or alkylmetallic compound.

Example:

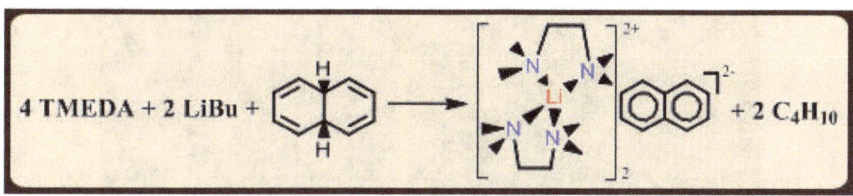

Sodium

Sodium, $_{11}$Na

Spectral lines of sodium

Sodium is a chemical element with symbol Na (from Latin *natrium*) and atomic number 11. It is a soft, silvery-white, highly reactive metal. Sodium is an alkali metal, being in group 1 of the periodic table, because it has a single electron in its outer shell that it readily donates, creating a positively charged atom—the Na^+ cation. Its only stable isotope is ^{23}Na. The free metal does not occur in nature, but must be prepared from compounds. Sodium is the sixth most abundant element in the Earth's crust, and exists in numerous minerals such as feldspars, sodalite and rock salt (NaCl). Many salts of sodium are highly water-soluble: sodium ions have been leached by the action of water from the Earth'sminerals over eons, and thus sodium and chlorine are the most common dissolved elements by weight in the oceans.

Sodium was first isolated by Humphry Davy in 1807 by the electrolysis of sodium hydroxide. Among many other useful sodium compounds, sodium hydroxide (lye) is used in soap manufacture, and sodium chloride (edible salt) is a de-icing agent and a nutrient for animals including humans.

Sodium is an essential element for all animals and some plants. Sodium ions are the major cation in the extracellular fluid (ECF) and as such are the major contributor to the ECF osmotic pressure and ECF compartment volume. Loss of water from the ECF compartment increases the sodium concentration, a condition called hypernatremia. Isotonic loss of water and sodium from the ECF compartment decreases the size of that compartment in a condition called ECF hypovolemia.

By means of the sodium-potassium pump, living human cells pump three sodium ions out of the cell in exchange for two potassium ions pumped in; comparing ion concentrations across the cell membrane, inside to outside, potassium measures about 40:1, and sodium, about 1:10. In nerve cells, the electrical charge across the cell membrane enables transmission of the nerve impulse—an action potential—when the charge is dissipated; sodium plays a key role in that activity.

Characteristics

Physical

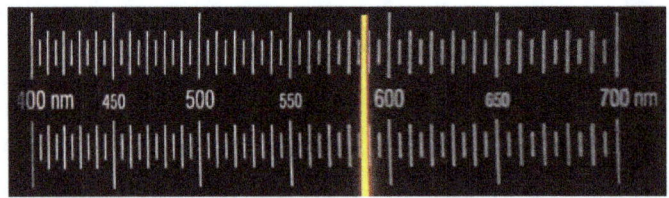

Emission spectrum for sodium, showing the D line.

Sodium at standard temperature and pressure is a soft silvery metal that combines with oxygen in air and forms grayish white sodium oxide unless immersed in oil or inert gas, which are the conditions it is usually stored in. Sodium metal can be easily cut with a knife and is a good conductor of electricity and heat because it has only one electron in its valence shell, resulting in weak metallic bonding and free electrons, which carry energy. Due to having low atomic mass and large atomic radius, sodium is third-least dense of all elemental metals and is one of only three metals that can float on water, the other two being lithium and potassium. The melting (98 °C) and boiling (883 °C) points of sodium are lower than those of lithium but higher than those of the heavier alkali metals potassium, rubidium, and caesium, following periodic trends down the group. These properties change dramatically at elevated pressures: at 1.5 Mbar, the color changes from silvery metallic to black; at 1.9 Mbar the material becomes transparent with a red color; and at 3 Mbar, sodium is a clear and transparent solid. All of these high-pressure allotropes are insulators and electrides.

A positive flame test for sodium has a bright yellow color.

In a flame test, sodium and its compounds glow yellow because the excited 3s electrons of sodium emit a photon when they fall from 3p to 3s; the wavelength of this photon corresponds to the D line at about 589.3 nm. Spin-orbit interactions involving the electron in the 3p orbital split the D line into two, at 589.0 and 589.6 nm; hyperfine structures involving both orbitals cause many more lines.

Isotopes

Twenty isotopes of sodium are known, but only ^{23}Na is stable. ^{23}Na is created in the carbon-burning process in stars by fusing two carbon atoms together; this requires temperatures above 600 megakelvins and a star of at least three solar masses. Two radioactive, cosmogenic isotopes are the byproduct of cosmic ray spallation: ^{22}Na has a half-life of 2.6 years and ^{24}Na, a half-life of 15 hours; all other isotopes have a half-life of less than one minute. Two nuclear isomers have been discovered, the longer-lived one being ^{24m}Na with a half-life of around 20.2 milliseconds. Acute neutron radiation, as from a nuclear criticality accident, converts some of the stable ^{23}Na in human blood to ^{24}Na; the neutron radiation dosage of a victim can be calculated by measuring the concentration of ^{24}Na relative to ^{23}Na.

Chemistry

Sodium atoms have 11 electrons, one more than the extremely stable configuration of the noble gasneon. Because of this and its low first ionization energy of 495.8 kJ/mol, the sodium atom is much more likely to lose the last electron and acquire a positive charge than to gain one and acquire a negative charge. This process requires so little energy that sodium is readily oxidized by giving up its 11th electron. In contrast, the second ionization energy is very high (4562 kJ/mol), because the 10th electron is closer to the nucleus than the 11th electron. As a result, sodium usually forms ionic compounds involving the Na^+ cation.

The most common oxidation state for sodium is +1. It is generally less reactive than potassium and more reactive than lithium. Sodium metal is highly reducing, with the reduction of sodium ions requiring −2.71 volts, though potassium and lithium have even more negative potentials.

Salts and Oxides

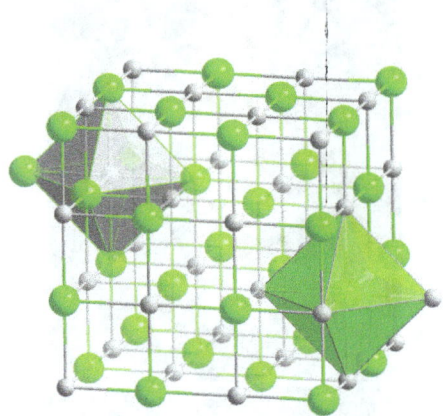

Structure of sodium chloride, showing octahedral coordination around Na^+ and Cl^- centres. This framework disintegrates when dissolved in water and reassembles when the water evaporates.

Sodium compounds are of immense commercial importance, being particularly central to industries producing glass, paper, soap, and textiles. The most important sodium compounds are table salt (NaCl), soda ash (Na_2CO_3), baking soda ($NaHCO_3$), caustic soda (NaOH), sodium nitrate ($NaNO_3$), di- and tri-sodium phosphates, sodium thiosulfate ($Na_2S_2O_3 \cdot 5H_2O$), and borax ($Na_2B_4O_7 \cdot 10H_2O$). In compounds, sodium is usually ionically bonded to water and anions, and is viewed as a hardLewis acid.

Two equivalent images of the chemical structure of sodium stearate, a typical soap.

Most soaps are sodium salts of fatty acids. Sodium soaps have a higher melting temperature (and seem "harder") than potassium soaps.

Like all the alkali metals, sodium reacts exothermically with water, and sufficiently large pieces melt to a sphere and may explode. The reaction produces caustic soda (sodium hydroxide) and flammable hydrogen gas. When burned in air, it forms primarily sodium peroxide with some sodium oxide.

Aqueous Solutions

Sodium tends to form water-soluble compounds, such as halides, sulfates, nitrates, carboxylates and carbonates. The main aqueous species are the aquo complexes $[Na(H_2O)_n]^+$, where $n = 4–8$; with $n = 6$ indicated from X-ray diffraction data and computer simulations.

Direct precipitation of sodium salts from aqueous solutions is rare because sodium salts typically have a high affinity for water; an exception is sodium bismuthate ($NaBiO_3$). Because of this, sodium salts are usually isolated as solids by evaporation or by precipitation with an organic solvent, such as ethanol; for example, only 0.35 g/L of sodium chloride will dissolve in ethanol. Crown ethers, like 15-crown-5, may be used as a phase-transfer catalyst.

Sodium content in bulk may be determined by treating with a large excess of uranyl zinc acetate; the hexahydrate $(UO_2)_2ZnNa(CH_3CO_2)\cdot6H_2O$ precipitates and can be weighed. Caesium and rubidium do not interfere with this reaction, but potassium and lithium do. Lower concentrations of sodium may be determined by atomic absorption spectrophotometry or by potentiometry using ion-selective electrodes.

Electrides and Sodides

Like the other alkali metals, sodium dissolves in ammonia and some amines to give deeply colored solutions; evaporation of these solutions leaves a shiny film of metallic sodium. The solutions contain the coordination complex $(Na(NH_3)_6)^+$, with the positive charge counterbalanced by electrons as anions; cryptands permit the isolation of these complexes as crystalline solids. Sodium forms complexes with crown ethers, cryptands and other ligands. For example, 15-crown-5 has high affinity for sodium because the cavity size of 15-crown-5 is 1.7–2.2 Å, which is enough to fit sodium ion (1.9 Å). Cryptands, like crown ethers and other ionophores, also have a high affinity for the sodium ion; derivatives of the alkalide Na^- are obtainable by the addition of cryptands to solutions of sodium in ammonia via disproportionation.

Organosodium Compounds

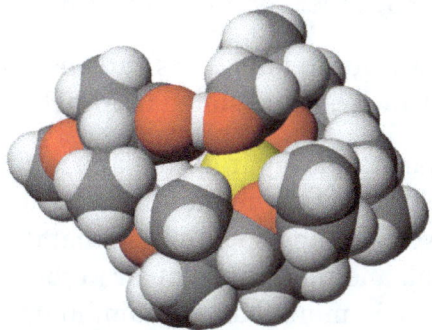

The structure of the complex of sodium (Na^+, shown in yellow) and the antibiotic monensin-A.

Many organosodium compounds have been prepared. Because of the high polarity of the C-Na bonds, they behave like sources of carbanions (salts with organic anions). Some well known derivatives include sodium cyclopentadienide (NaC_5H_5) and trityl sodium ($(C_6H_5)_3CNa$). Because of the large size and very low polarising power of the Na^+ cation, it can stabilize large, aromatic, polarisable radical anions, such as in sodium naphthalenide, $Na^+[C_{10}H_8\bullet]^-$, a strong reducing agent.

Intermetallic Compounds

Sodium forms alloys with many metals, such as potassium, calcium, lead, and the group 11 and 12 elements. Sodium and potassium form KNa_2 and NaK. NaK is 40–90% potassium and it is liquid at ambient temperature. It is excellent thermal and electrical conductor. Sodium-calcium alloys are by-products of electrolytic production of sodium from binary salt mixture of $NaCl$-$CaCl_2$ and ternary mixture $NaCl$-$CaCl_2$-$BaCl_2$. Calcium is only partially miscible with sodium. In liquid state, sodium is completely miscible with lead. There are several methods to make sodium-lead alloys. One is to melt them together and another is to deposit sodium electrolycally on molten lead cathodes. $NaPb_3$, $NaPb$, Na_9Pb_4, Na_5Pb_2, and $Na_{15}Pb_4$ are some of the known sodium-lead alloys. Sodium also forms alloys with gold ($NaAu_2$) and silver ($NaAg_2$). Group 12 metals (zinc, cadmium and mercury) are known to make alloys with sodium. $NaZn_{13}$ and $NaCd_2$ are alloys of zinc and cadmium. Sodium and mercury form $NaHg$, $NaHg_4$, $NaHg_2$, Na_3Hg_2, and Na_3Hg.

History

Because of its importance in human metabolism, salt has long been an important commodity as shown by the English word *salary*, which derives from *salarium*, the wafers of salt sometimes given to Roman soldiers along with their other wages. In medieval Europe, a compound of sodium with the Latin name of *sodanum* was used as a headache remedy. The name sodium is thought to originate from the Arabic *suda*, meaning headache, as the headache-alleviating properties of sodium carbonate or soda were well known in early times. Although sodium, sometimes called *soda*, had long been recognized in compounds, the metal itself was not isolated until 1807 by Sir Humphry Davy through the electrolysis of sodium hydroxide. In 1809, the German physicist and chemist Ludwig Wilhelm Gilbert proposed the names *Natronium* for Humphry Davy's "sodium" and *Kalium* for Davy's "potassium". The chemical abbreviation for sodium was first published in 1814 by Jöns Jakob Berzelius in his system of atomic symbols, and is an abbreviation of the element's New Latin name *natrium*, which refers to the Egyptian *natron*, a natural mineral salt mainly consisting of hydrated sodium carbonate. Natron historically had several important industrial and household uses, later eclipsed by other sodium compounds.

Sodium imparts an intense yellow color to flames. As early as 1860, Kirchhoff and Bunsen noted the high sensitivity of a sodium flame test, and stated in Annalen der Physik und Chemie:

In a corner of our 60 m³ room farthest away from the apparatus, we exploded 3 mg. of sodium chlorate with milk sugar while observing the nonluminous flame before the slit. After a while, it glowed a bright yellow and showed a strong sodium line that disappeared only after 10 minutes. From the weight of the sodium salt and the volume of air in the room, we easily calculate that one part by weight of air could not contain more than 1/20 millionth weight of sodium.

Occurrence

The Earth's crust contains 2.27% sodium, making it the seventh most abundant element on Earth and the fifth most abundant metal, behind aluminium, iron, calcium, and magnesium and ahead of potassium. Sodium's estimated oceanic abundance is 1.08×10^4 milligrams per liter. Because of its high reactivity, it is never found as a pure element. It is found in many different minerals, some very soluble, such as halite and natron, others much less soluble, such as amphibole and zeolite. The insolubility of certain sodium minerals such as cryolite and feldspar arises from their polymeric anions, which in the case of feldspar is a polysilicate. In the interstellar medium, sodium is identified by the D spectral line; though it has a high vaporization temperature, its abundance in Mercury's atmosphere enabled its detection by Potter and Morgan using ground-based high resolution spectroscopy. Sodium has been detected in at least one comet; astronomers watching Comet Hale-Bopp in 1997 observed a sodium tail consisting of neutral atoms (not ions) and extending to some 50 million kilometres behind the head.

Commercial Production

Employed only in rather specialized applications, only about 100,000 tonnes of metallic sodium are produced annually. Metallic sodium was first produced commercially in the late 19th century by carbothermal reduction of sodium carbonate at 1100 °C, as the first step of the Deville process for the production of aluminium:

$$Na_2CO_3 + 2\,C \rightarrow 2\,Na + 3\,CO$$

The high demand of aluminium created the need for the production of sodium. After the introduction of the Hall–Héroult process for the production of aluminium in by electrolysing a molten salt bath ended the need for large quantities of sodium. A related process based on the reduction of sodium hydroxide was developed in 1886.

Sodium is now produced commercially through the electrolysis of molten sodium chloride, based on a process patented in 1924. This is done in a Downs cell in which the NaCl is mixed with calcium chloride to lower the melting point below 700 °C. As calcium is less electropositive than sodium, no calcium will be deposited at the cathode. This method is less expensive than the previous Castner process (the electrolysis of sodium hydroxide).

The market for sodium is volatile due to the difficulty in its storage and shipping; it must be stored under a dry inert gas atmosphere or anhydrousmineral oil to prevent the formation of a surface layer of sodium oxide or sodium superoxide.

Applications

Though metallic sodium has some important uses, the major applications for sodium use compounds; millions of tons of sodium chloride, hydroxide, and carbonate are produced annually. Sodium chloride is extensively used for anti-icing and de-icing and as a preservative; sodium bicarbonate is mainly used for cooking. Along with potassium, many important medicines have sodium added to improve their bioavailability; though potassium is the better ion in most cases, sodium is chosen for its lower price and atomic weight. Sodium hydride is used as a base for various reactions (such as the aldol reaction) in organic chemistry, and as a reducing agent in inorganic chemistry.

Free Element

Metallic sodium is used mainly for the production of sodium borohydride, sodium azide, indigo, and triphenylphosphine. Previous uses were for the making of tetraethyllead and titanium metal; because applications for these chemicals were discontinued, the production of sodium declined after 1970. Sodium is also used as an alloying metal, an anti-scaling agent, and as a reducing agent for metals when other materials are ineffective. Note the free element is not used as a scaling agent, ions in the water are exchanged for sodium ions. Sodium plasma ("vapor") lamps are often used for street lighting in cities, shedding light that ranges from yellow-orange to peach as the pressure increases. By itself or with potassium, sodium is a desiccant; it gives an intense blue coloration with benzophenone when the desiccate is dry. In organic synthesis, sodium is used in various reactions such as the Birch reduction, and the sodium fusion test is conducted to qualitatively analyse compounds. Sodium reacts with alcohol and gives alkoxides, and when sodium is dissolved in ammonia solution, it can be used to reduce alkynes to trans-alkenes. Sodium lasers emitting light at the D line are used to create artificial laser guide stars that assist in the adaptive optics for land-based visible light telescopes.

Heat Transfer

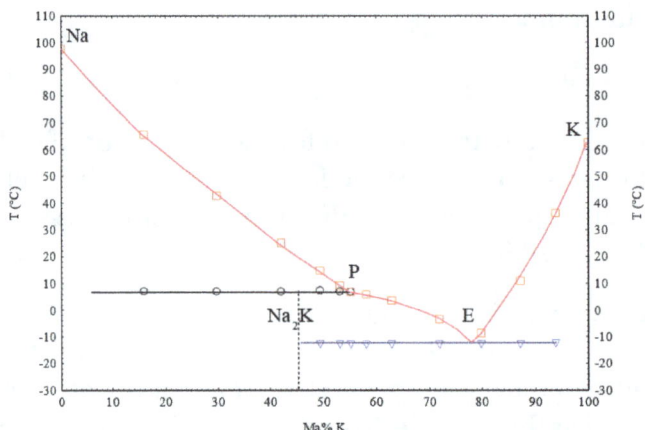

NaK phase diagram, showing the melting point of sodium as a function of potassium concentration. NaK with 77% potassium is eutectic and has the lowest melting point of the NaK alloys at −12.6 °C.

Liquid sodium is used as a heat transfer fluid in some fast reactors because it has the high thermal conductivity and low neutron absorption cross section required to achieve a high neutron flux in the reactor. The high boiling point of sodium allows the reactor to operate at ambient (normal) pressure, but the drawbacks include its opacity, which hinders visual maintenance, and its explosive properties. Radioactive sodium-24 may be produced by neutron bombardment during operation, posing a slight radiation hazard; the radioactivity stops within a few days after removal from the reactor. If a reactor needs to be shut down frequently, NaK is used; because NaK is a liquid at room temperature, the coolant does not solidify in the pipes. In this case, the pyrophoricity of potassium requires extra precautions to prevent and detect leaks. Another heat transfer application is poppet valves in high-performance internal combustion engines; the valve stems are partially filled with sodium and work as a heat pipe to cool the valves.

Biological Role

In humans, sodium is an essential mineral that regulates blood volume, blood pressure, osmotic equilibrium and pH; the minimum physiological requirement for sodium is 500 milligrams per day.Sodium chloride is the principal source of sodium in the diet, and is used as seasoning and preservative in such commodities as pickled preserves and jerky; for Americans, most sodium chloride comes from processed foods. Other sources of sodium are its natural occurrence in food and such food additives as monosodium glutamate (MSG), sodium nitrite, sodium saccharin, baking soda (sodium bicarbonate), and sodium benzoate. The US Institute of Medicine set its Tolerable Upper Intake Level for sodium at 2.3 grams per day, but the average person in the United States consumes 3.4 grams per day. Studies have found that lowering sodium intake by 2 g per day tends to lower systolic blood pressure by about two to four mm Hg. It has been estimated that such a decrease in sodium intake would lead to between 9 and 17% fewer cases of hypertension.

Hypertension causes 7.6 million premature deaths worldwide each year. (Note that salt contains about 39.3% sodium—the rest being chlorine and trace chemicals; thus, 2.3 g sodium is about 5.9 g, or 2.7 ml of salt—about half a US teaspoon.) The American Heart Association recommends no more than 1.5 g of sodium per day.

One study found that people with or without hypertension who excreted less than 3 grams of sodium per day in their urine (and therefore were taking in less than 3 g/d) had a *higher* risk of death, stroke, or heart attack than those excreting 4 to 5 grams per day. Levels of 7 g per day or more in people with hypertension were associated with higher mortality and cardiovascular events, but this was not found to be true for people without hypertension. The US FDA states that adults with hypertension and prehypertension should reduce daily intake to 1.5 g.

The renin-angiotensin system regulates the amount of fluid and sodium concentration in the body. Reduction of blood pressure and sodium concentration in the kidney result in the production of renin, which in turn produces aldosterone and angiotensin, retaining sodium in the urine. When the concentration of sodium increases, the production of renin decreases, and the sodium concentration returns to normal. The sodium ion (Na^+) is an important electrolyte in neuron function, and in osmoregulation between cells and the extracellular fluid. This is accomplished in all animals by Na^+/K^+-ATPase, an active transporter pumping ions against the gradient, and sodium/potassium channels. Sodium is the most prevalent metallic ion in extracellular fluid.

Unusually low or high sodium levels in humans are recognized in medicine as hyponatremia and hypernatremia. These conditions may be caused by genetic factors, ageing, or prolonged vomiting or diarrhea.

In C4 plants, sodium is a micronutrient that aids in metabolism, specifically in regeneration of phosphoenolpyruvate and synthesis of chlorophyll. In others, it substitutes for potassium in several roles, such as maintaining turgor pressure and aiding in the opening and closing of stomata. Excess sodium in the soil limits the uptake of water by decreasing the water potential, which may result in plant wilting; excess concentrations in the cytoplasm can lead to enzyme inhibition, which in turn causes necrosis and chlorosis. In response, some plants developed mechanisms to limit sodium uptake in the roots, to store it in cell vacuoles, and restriction of salt transport from roots to leaves; excess sodium may also be stored in old plant tissue, limiting the damage to new growth.

Safety and Precautions

Sodium forms flammable hydrogen and caustic sodium hydroxide on contact with water; ingestion and contact with moisture on skin, eyes or mucous membranes can cause severe burns. Sodium spontaneously explodes in the presence of an oxidizer such as water.Fire extinguishers based on water accelerate sodium fires; those based on carbon dioxide and bromochlorodifluoromethane should not be used on sodium fire. Metal fires are Class D, but not all Class D extinguishers are workable with sodium. An effective extinguishing agent for sodium fires is Met-L-X. Other effective agents include Lith-X, which has graphite powder and an organophosphateflame retardant, and dry sand. Sodium fires are prevented in nuclear reactors by isolating sodium from oxygen by surrounding sodium pipes with inert gas. Pool-type sodium fires are prevented using different design measures called catch pan systems. They collect leaking sodium into a leak-recovery tank where it is isolated from oxygen.

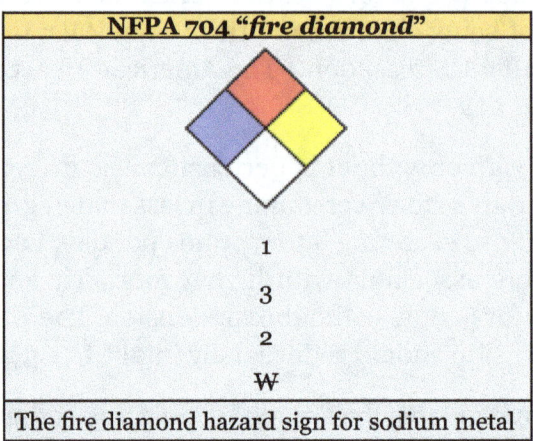

The fire diamond hazard sign for sodium metal

Lithium

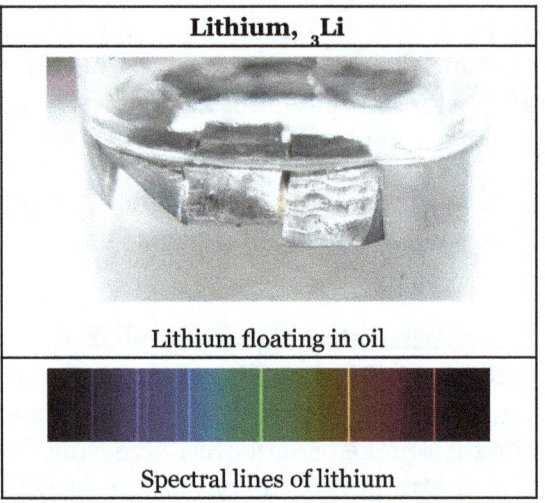

Lithium floating in oil

Spectral lines of lithium

Lithium (from Greek: *lithos*, "stone") is a chemical element with the symbol Li and atomic number 3. It is a soft, silver-white metal belonging to the alkali metalgroup of chemical elements. Under standard conditions, it is the lightest metal and the least dense solid element. Like all alkali metals, lithium is highly reactive and flammable. For this reason, it is typically stored in mineral oil. When cut open, it exhibits a metallic luster, but contact with moist air corrodes the surface quickly to a dull silvery gray, then black tarnish. Because of its high reactivity, lithium never occurs freely in nature, and instead, appears only in compounds, which are usually ionic. Lithium occurs in a number of pegmatitic minerals, but due to its solubility as an ion, is present in ocean water and is commonly obtained from brines and clays. On a commercial scale, lithium is isolated electrolytically from a mixture of lithium chloride and potassium chloride.

The nucleus of the lithium atom verges on instability, since the two stable lithium isotopes found in nature have among the lowest binding energies per nucleon of all stable nuclides. Because of its relative nuclear instability, lithium is less common in the solar system than 25 of the first 32 chemical elements even though the nuclei are very light in atomic weight. For related reasons, lithium has important links to nuclear physics. The transmutation of lithium atoms to helium in 1932 was the first fully man-made nuclear reaction, and lithium-6 deuteride serves as a fusion fuel in staged thermonuclear weapons.

Lithium and its compounds have several industrial applications, including heat-resistant glass and ceramics, lithium grease lubricants, flux additives for iron, steel and aluminium production, lithium batteries, and lithium-ion batteries. These uses consume more than three quarters of lithium production.

Lithium is found in variable amounts in foods; primary food sources are grains and vegetables; in some areas, the drinking water also provides significant amounts of the element. Human dietary lithium intakes depend on location and the type of foods consumed and vary over a wide range. Traces of lithium were detected in human organs and fetal tissues already in the late 19th century, leading to early suggestions as to possible specific functions in the organism. However, it took another century until evidence for the essentiality of lithium became available. In studies conducted from the 1970s to the 1990s, rats and goats maintained on low-lithium rations were shown to exhibit higher mortalities as well as reproductive and behavioral abnormalities. In humans defined lithium deficiency diseases have not been characterized, but low lithium intakes from water supplies were associated with increased rates of suicides, homicides and the arrest rates for drug use and other crimes. Lithium appears to play an especially important role during the early fetal development as evidenced by the high lithium contents of the embryo during the early gestational period. The biochemical mechanisms of action of lithium appear to be multifactorial and are inter-correlated with the functions of several enzymes, hormones and vitamins, as well as with growth and transforming factors. The available experimental evidence now appears to be sufficient to accept lithium as essential; a provisional RDA for a 70 kg adult of 1,000 µg/day is suggested.

The lithium ion Li$^+$ administered as any of several lithium salts has proven to be useful as a mood-stabilizing drug in the treatment of bipolar disorder in humans.

Properties

Atomic and Physical

Lithium ingots with a thin layer of black nitride tarnish

Like the other alkali metals, lithium has a single valence electron that is easily given up to form a cation. Because of this, lithium is a good conductor of heat and electricity as well as a highly reactive element, though it is the least reactive of the alkali metals. Lithium's low reactivity is due to the proximity of its valence electron to its nucleus (the remaining two electrons are in the 1s orbital, much lower in energy, and do not participate in chemical bonds).

Lithium metal is soft enough to be cut with a knife. When cut, it possesses a silvery-white color that quickly changes to gray as it oxidizes to lithium oxide. While it has one of the lowest melting points among all metals (180 °C), it has the highest melting and boiling points of the alkali metals.

Lithium has a very low density (0.534 g/cm³), comparable with pine wood. It is the least dense of all elements that are solids at room temperature; the next lightest solid element (potassium, at 0.862 g/cm³) is more than 60% denser. Furthermore, apart from helium and hydrogen, it is less dense than any liquid element, being only two thirds as dense as liquid nitrogen (0.808 g/cm³). Lithium can float on the lightest hydrocarbon oils and is one of only three metals that can float on water, the other two being sodium and potassium.

Lithium floating in oil

Lithium's coefficient of thermal expansion is twice that of aluminium and almost four times that of iron. Lithium is superconductive below 400 μK at standard pressure and at higher temperatures (more than 9 K) at very high pressures (>20 GPa). At temperatures below 70 K, lithium, like sodium, undergoes diffusionless phase change transformations. At 4.2 K it has a rhombohedral crystal system (with a nine-layer repeat spacing); at higher temperatures it transforms to face-centered cubic and then body-centered cubic. At liquid-helium temperatures (4 K) the rhombohedral structure is prevalent. Multiple allotropic forms have been identified for lithium at high pressures.

Lithium has a mass specific heat capacity of 3.58 kilojoules per kilogram-kelvin, the highest of all solids. Because of this, lithium metal is often used in coolants for heat transfer applications.

Chemistry and Compounds

Lithium reacts with water easily, but with noticeably less energy than other alkali metals. The reaction forms hydrogen gas and lithium hydroxide in aqueous solution. Because of its reactivity with water, lithium is usually stored in a hydrocarbon sealant, often petroleum jelly. Though the heavier alkali metals can be stored in more dense substances, such as mineral oil, lithium is not dense enough to be fully submerged in these liquids. In moist air, lithium rapidly tarnishes to form a black coating of lithium hydroxide (LiOH and LiOH·H$_2$O), lithium nitride (Li$_3$N) and lithium carbonate (Li$_2$CO$_3$, the result of a secondary reaction between LiOH and CO$_2$).

Hexameric structure of the n-butyllithium fragment in a crystal

When placed over a flame, lithium compounds give off a striking crimson color, but when it burns strongly the flame becomes a brilliant silver. Lithium will ignite and burn in oxygen when exposed to water or water vapors. Lithium is flammable, and it is potentially explosive when exposed to air and especially to water, though less so than the other alkali metals. The lithium-water reaction at normal temperatures is brisk but nonviolent because the hydrogen produced does not ignite on its own. As with all alkali metals, lithium fires are difficult to extinguish, requiring dry powder fire extinguishers (Class D type). Lithium is the only metal which reacts with nitrogen under normal conditions.

Lithium has a diagonal relationship with magnesium, an element of similar atomic and ionic radius. Chemical resemblances between the two metals include the formation of a nitride by reaction with N$_2$, the formation of an oxide (Li$_2$O) and peroxide (Li$_2$O$_2$) when burnt in O$_2$, salts with similar solubilities, and thermal instability of the carbonates and nitrides. The metal reacts with hydrogen gas at high temperatures to produce lithium hydride (LiH).

Other known binary compounds include halides (LiF, LiCl, LiBr, LiI), sulfide (Li$_2$S), superoxide (LiO$_2$), and carbide (Li2C$_2$). Many other inorganic compounds are known in which lithium combines with anions to form salts: borates, amides, carbonate, nitrate, or borohydride (LiBH$_4$). Lithium aluminium hydride (LiAlH$_4$) is commonly used as a reducing agent in organic synthesis.

Multiple organolithium reagents are known in which there is a direct bond between carbon and lithium atoms, effectively creating a carbanion. These are extremely powerful bases and nucleophiles. In many of these organolithium compounds, the lithium ions tend to aggregate into high-symmetry clusters by themselves, which is relatively common for alkali cations.LiHe, a very weakly interacting van der Waals compound, has been detected at very low temperatures.

Isotopes

Naturally occurring lithium is composed of two stable isotopes, ^{6}Li and ^{7}Li, the latter being the more abundant (92.5% natural abundance). Both natural isotopes have anomalously low nuclear binding energy per nucleon (compared to the neighboring elements on the periodic table, helium and beryllium); lithium is the only low numbered element that can produce net energy through nuclear fission. The two lithium nuclei have lower binding energies per nucleon than any other stable nuclides other than deuterium and helium-3. As a result of this, though very light in atomic weight, lithium is less common in the Solar System than 25 of the first 32 chemical elements. Seven radioisotopes have been characterized, the most stable being ^{8}Li with a half-life of 838 ms and ^{9}Li with a half-life of 178 ms. All of the remaining radioactive isotopes have half-lives that are shorter than 8.6 ms. The shortest-lived isotope of lithium is ^{4}Li, which decays through proton emission and has a half-life of 7.6×10^{-23} s.

^{7}Li is one of the primordial elements (or, more properly, primordial nuclides) produced in Big Bang nucleosynthesis. A small amount of both ^{6}Li and ^{7}Li are produced in stars, but are thought to be "burned" as fast as produced. Additional small amounts of lithium of both ^{6}Li and ^{7}Li may be generated from solar wind, cosmic rays hitting heavier atoms, and from early solar system ^{7}Be and ^{10}Be radioactive decay. While lithium is created in stars during stellar nucleosynthesis, it is further burned. ^{7}Li can also be generated in carbon stars.

Lithium isotopes fractionate substantially during a wide variety of natural processes, including mineral formation (chemical precipitation), metabolism, and ion exchange. Lithium ions substitute for magnesium and iron in octahedral sites in clay minerals, where ^{6}Li is preferred to ^{7}Li, resulting in enrichment of the light isotope in processes of hyperfiltration and rock alteration. The exotic ^{11}Li is known to exhibit a nuclear halo. The process known as laser isotope separation can be used to separate lithium isotopes, in particular ^{7}Li from ^{6}Li.

Nuclear weapons manufacture and other nuclear physics applications are a major source of artificial lithium fractionation, with the light isotope ^{6}Li being retained by industry and military stockpiles to such an extent that it has caused slight but measurable change in the ^{6}Li to ^{7}Li ratios in natural sources, such as rivers. This has led to unusual uncertainty in the standardized atomic weight of lithium, since this quantity depends on the natural abundance ratios of these naturally-occurring stable lithium isotopes, as they are available in commercial lithium mineral sources.

Occurrence

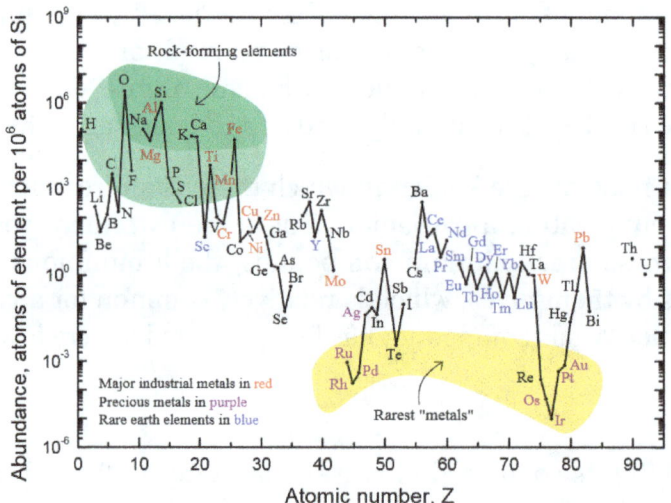

Lithium is about as common as chlorine in the Earth's upper continental crust, on a per-atom basis.

Astronomical

According to modern cosmological theory, lithium—in both stable isotopes (lithium-6 and lithium-7)—was one of the 3 elements synthesized in the Big Bang. Though the amount of lithium generated in Big Bang nucleosynthesis is dependent upon the number of photons per baryon, for accepted values the lithium abundance can be calculated, and there is a "cosmological lithium discrepancy" in the Universe: older stars seem to have less lithium than they should, and some younger stars have much more. The lack of lithium in older stars is apparently caused by the "mixing" of lithium into the interior of stars, where it is destroyed, while lithium is produced in younger stars. Though it transmutes into two atoms of helium due to collision with a proton at temperatures above 2.4 million degrees Celsius (most stars easily attain this temperature in their interiors), lithium is more abundant than current computations would predict in later-generation stars.

Nova Centauri 2013 is the first in which evidence of lithium has been found.

Though it was one of the three first elements to be synthesized in the Big Bang, lithium, together with beryllium and boron are markedly less abundant than other elements. This is a result of the low temperature necessary to destroy lithium, and a lack of common processes to produce it.

Lithium is also found in brown dwarf substellar objects and certain anomalous orange stars. Because lithium is present in cooler, less-massive brown dwarfs, but is destroyed in hotter red dwarf stars, its presence in the stars' spectra can be used in the "lithium test" to differentiate the two, as both are smaller than the Sun. Certain orange stars can also contain a high concentration of lithium. Those orange stars found to have a higher than usual concentration of lithium (such as Centaurus X-4) orbit massive objects—neutron stars or black holes—whose gravity evidently pulls heavier lithium to the surface of a hydrogen-helium star, causing more lithium to be observed.

Terrestrial

Lithium mine production (2015) and reserves in tonnes		
Country	**Production**	**Reserves**
Argentina	3,800	2,000,000
Australia	13,400	1,500,000
Brazil	160	48,000
Canada (2010)	480	180,000
Chile	11,700	7,500,000
People's Republic of China	2,200	3,200,000
Portugal	300	60,000
Zimbabwe	900	23,000
World total	**32,500**	**14,000,000**

Although lithium is widely distributed on Earth, it does not naturally occur in elemental form due to its high reactivity. The total lithium content of seawater is very large and is estimated as 230 billion tonnes, where the element exists at a relatively constant concentration of 0.14 to 0.25 parts per million (ppm), or 25 micromolar; higher concentrations approaching 7 ppm are found near hydrothermal vents.

Estimates for the Earth's crustal content range from 20 to 70 ppm by weight. In keeping with its name, lithium forms a minor part of igneous rocks, with the largest concentrations in granites. Granitic pegmatites also provide the greatest abundance of lithium-containing minerals, with spodumene and petalite being the most commercially viable sources. Another significant mineral of lithium is lepidolite. A newer source for lithium is hectorite clay, the only active development of which is through the Western Lithium Corporation in the United States. At 20 mg lithium per kg of Earth's crust, lithium is the 25th most abundant element.

According to the *Handbook of Lithium and Natural Calcium*, "Lithium is a comparatively rare element, although it is found in many rocks and some brines, but always in very low concentrations. There are a fairly large number of both lithium mineral and brine deposits but only comparatively few of them are of actual or potential commercial value. Many are very small, others are too low in grade."

The US Geological Survey estimates that in 2010, Chile had the largest reserves by far (7.5 million tonnes) and the highest annual production (8,800 tonnes). One of the largest *reserve bases* of lithium is in the Salar de Uyuni area of Bolivia, which has 5.4 million tonnes. Other major suppliers include Australia, Argentina and China. As of 2015 Czech Geological Survey considered the entire Ore Mountains in the Czech Republic as lithium province. Five deposits are registered, one near Cínovec (cs) is considered as potentially economic deposit with 160 000 tonnes of lithium.

In June 2010, the *New York Times* reported that American geologists were conducting ground surveys on drysalt lakes in western Afghanistan believing that large deposits of lithium are located there. "Pentagon officials said that their initial analysis at one location in Ghazni Province showed the potential for lithium deposits as large as those of Bolivia, which now has the world's largest known lithium reserves." These estimates are "based principally on old data, which was gathered mainly by the Soviets during their occupation of Afghanistan from 1979–1989". Stephen Peters, the head of the USGS's Afghanistan Minerals Project, said that he was unaware of USGS involvement in any new surveying for minerals in Afghanistan in the past two years. 'We are not aware of any discoveries of lithium,' he said."

Lithia ("lithium brine") is associated with tin mining areas in Cornwall, England and an evaluation project from 400-metre deep test boreholes is under consideration. If successful the hot brines will also provide geothermal energy to power the lithium extraction and refining process.

Biological

Lithium is found in trace amount in numerous plants, plankton, and invertebrates, at concentrations of 69 to 5,760 parts per billion (ppb). In vertebrates the concentration is slightly lower, and nearly all vertebrate tissue and body fluids contain lithium ranging from 21 to 763 ppb. Marine organisms tend to bioaccumulate lithium more than terrestrial organisms. Whether lithium has a physiological role in any of these organisms is unknown.

History

Johan August Arfwedson is credited with the discovery of lithium in 1817

Petalite ($LiAlSi_4O_{10}$) was discovered in 1800 by the Brazilian chemist and statesman José Bonifácio de Andrada e Silva in a mine on the island of Utö, Sweden. However, it was not until 1817 that Johan August Arfwedson, then working in the laboratory of the chemist Jöns Jakob Berzelius, detected the presence of a new element while analyzing petalite ore. This element formed compounds similar to those of sodium and potassium, though its carbonate and hydroxide were less soluble in water and more alkaline. Berzelius gave the alkaline material the name "*lithion/lithina*", from the Greekword (*lithos*, meaning "stone"), to reflect its discovery in a solid mineral, as opposed to potassium, which had been discovered in plant ashes, and sodium which was known partly for its high abundance in animal blood. He named the metal inside the material "*lithium*".

Arfwedson later showed that this same element was present in the minerals spodumene and lepidolite. In 1818, Christian Gmelin was the first to observe that lithium salts give a bright red color to flame. However, both Arfwedson and Gmelin tried and failed to isolate the pure element from its salts. It was not isolated until 1821, when William Thomas Brande obtained it by electrolysis of lithium oxide, a process that had previously been employed by the chemist Sir Humphry Davy to isolate the alkali metals potassium and sodium. Brande also described some pure salts of lithium, such as the chloride, and, estimating that lithia (lithium oxide) contained about 55% metal, estimated the atomic weight of lithium to be around 9.8 g/mol (modern value ~6.94 g/mol). In 1855, larger quantities of lithium were produced through the electrolysis of lithium chloride by Robert Bunsen and Augustus Matthiessen. The discovery of this procedure led to commercial production of lithium in 1923 by the German company Metallgesellschaft AG, which performed an electrolysis of a liquid mixture of lithium chloride and potassium chloride.

The production and use of lithium underwent several drastic changes in history. The first major application of lithium was in high-temperature lithium greases for aircraft engines and similar applications in World War II and shortly after. This use was supported by the fact that lithium-based soaps have a higher melting point than other alkali soaps, and are less corrosive than calcium based soaps. The small market for lithium soaps and lubricating greases was supported by several small mining operations mostly in the United States.

The demand for lithium increased dramatically during the Cold War with the production of nuclear fusion weapons. Both lithium-6 and lithium-7 produce tritium when irradiated by neutrons, and are thus useful for the production of tritium by itself, as well as a form of solid fusion fuel used inside hydrogen bombs in the form of lithium deuteride. The United States became the prime producer of lithium in the period between the late 1950s and the mid-1980s. At the end, the stockpile of lithium was roughly 42,000 tonnes of lithium hydroxide. The stockpiled lithium was depleted in lithium-6 by 75%, which was enough to affect the measured atomic weight of lithium in many standardized chemicals, and even the atomic weight of lithium in some "natural sources" of lithium ion which had been "contaminated" by lithium salts discharged from isotope separation facilities, which had found its way into ground water.

Lithium was used to decrease the melting temperature of glass and to improve the melting behavior of aluminium oxide when using the Hall-Héroult process. These two uses dominated the market until the middle of the 1990s. After the end of the nuclear arms race, the demand for lithium decreased and the sale of Department of Energy stockpiles on the open market further reduced prices. But in the mid-1990s, several companies started to extract lithium from brine which proved to be a less expensive method than underground or even open-pit mining. Most of the mines closed or shifted their focus to other materials because only the ore from zoned pegmatites could be mined for a competitive price. For example, the US mines near Kings Mountain, North Carolina closed before the turn of the 21st century.

The development of lithium ion batteries increased the demand for lithium and became the dominant use in 2007. With the surge of lithium demand in batteries in the 2000s, new companies have expanded brine extraction efforts to meet the rising demand.

Production

Satellite images of the Salar del Hombre Muerto, Argentina (left), and Uyuni, Bolivia (right), salt flats that are rich in lithium. The lithium-rich brine is concentrated by pumping it into solar evaporation ponds.

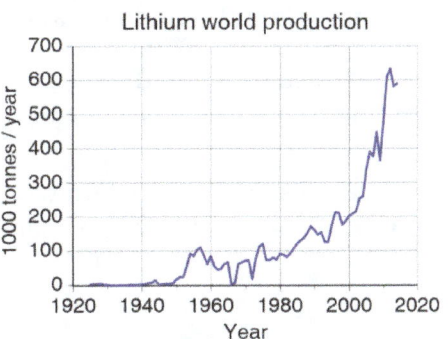

World production trend of lithium

Lithium production has greatly increased since the end of World War II. The metal is separated from other elements in igneous minerals. Lithium salts are extracted from water in mineral springs, brine pools, and brine deposits. The metal is produced through electrolysis from a mixture of fused 55% lithium chloride and 45% potassium chloride at about 450 °C.

In 1998, the price of lithium was about 95 USD/kg (or 43 USD/lb).

Reserves

Worldwide identified reserves in 2008 were estimated by the US Geological Survey (USGS) to be 13 million tonnes, though an accurate estimate of world lithium reserves is difficult.

Deposits are found in South America throughout the Andes mountain chain. Chile is the leading producer, followed by Argentina. Both countries recover lithium from brine pools. In the United States, lithium is recovered from brine pools in Nevada. However, half the world's known reserves are located in Bolivia along the central eastern slope of the Andes. In 2009, Bolivia negotiated with Japanese, French, and Korean firms to begin extraction. According to USGS, Bolivia's Uyuni Desert has 5.4 million tonnes of lithium. A newly discovered deposit in Wyoming's Rock Springs Uplift is estimated to contain 228,000 tons. Additional deposits in the same formation were estimated to be as much as 18 million tons.

Opinions differ about potential growth. A 2008 study concluded that "realistically achievable lithium carbonate production will be sufficient for only a small fraction of future PHEV and EV global market requirements", that "demand from the portable electronics sector will absorb much of the planned production increases in the next decade", and that "mass production of lithium carbonate is not environmentally sound, it will cause irreparable ecological damage to ecosystems that should be protected and that LiIon propulsion is incompatible with the notion of the 'Green Car'".

However, according to a 2011 study conducted at Lawrence Berkeley National Laboratory and the University of California, Berkeley, the currently estimated reserve base of lithium should not be a limiting factor for large-scale battery production for electric vehicles because an estimated 1 billion 40 kWh Li-based batteries could be built with current reserves - about 10 kg of lithium per car. Another 2011 study by researchers from the University of Michigan and Ford Motor Company found sufficient resources to support global demand until 2100, including the lithium required for the potential widespread transportation use. The study estimated global reserves at 39 million tons, and total demand for lithium during the 90-year period analyzed at 12–20 million tons, depending on the scenarios regarding economic growth and recycling rates.

On June 9, 2014, the *Financialist* stated that demand for lithium was growing at more than 12 percent a year; according to Credit Suisse, this rate exceeds projected availability by 25 percent. The publication compared the 2014 lithium situation with oil, whereby "higher oil prices spurred investment in expensive deepwater and oil sands production techniques"; that is, the price of lithium will continue to rise until more expensive production methods that can boost total output receive the attention of investors.

Pricing

After the 2007 financial crisis, major suppliers such as Sociedad Química y Minera (SQM) dropped lithium carbonate pricing by 20%. Prices rose in 2012. A 2012 Business Week article outlined the oligopoly in the lithium space: "SQM, controlled by billionaire Julio Ponce, is the second-largest, followed by Rockwood, which is backed by Henry Kravis's KKR & Co., and Philadelphia-based FMC". Global consumption may jump to 300,000 metric tons a year by 2020 from about 150,000 tons in 2012, to match the demand for lithium batteries that has been growing at about 25 percent a year, outpacing the 4 percent to 5 percent overall gain in lithium production.

Extraction

As of 2015 most of the world's lithium production is in South America, where lithium-containing brine is extracted from underground pools and concentrated by solar evaporation. The standard extraction technique is to evaporate water from brine. Each batch takes from 18 to 24 months.

Seawater

Lithium is present in seawater, but commercially viable methods of extraction have yet to be developed.

Geothermal Wells

One potential source of lithium is the leachates of geothermal wells, which are carried to the surface. recovery of lithium has been demonstrated in the field. The lithium is separated by simple filtration. The process and environmental costs are primarily those of the already-operating well; net environmental impacts may thus be positive.

Uses

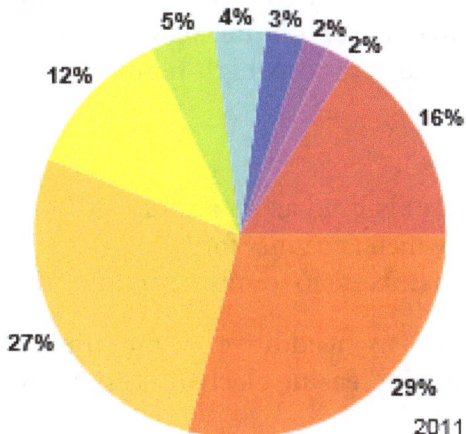

Estimates of global lithium uses in 2015

- Ceramics and glass (32%)
- Batteries (35%)
- Lubricating greases (9%)
- Continuous casting (5%)
- Air treatment (5%)
- Polymers (4%)
- Primary aluminum production (1%)
- Pharmaceuticals (<1%)
- Other (9%)

Ceramics and Glass

Lithium oxide is widely used as a flux for processing silica, reducing the melting point and viscosity of the material and leading to glazes with improved physical properties including low coefficients of thermal expansion. Worldwide, this is the single largest use for lithium compounds. Glazes containing lithium oxides are used for ovenware. Lithium carbonate (Li_2CO_3) is generally used in this application because it converts to the oxide upon heating.

Electrical and Electronics

Late in the 20th century, lithium became an important component of battery electrolytes and electrodes, because of its high electrode potential. Because of its low atomic mass, it has a high charge- and power-to-weight ratio. A typical lithium-ion battery can generate approximately 3 volts per cell, compared with 2.1 volts for lead-acid or 1.5 volts for zinc-carbon cells. Lithium-ion batteries, which are rechargeable and have a high energy density, should not be confused with lithium batteries, which are disposable (primary) batteries with lithium or its compounds as the anode. Other rechargeable batteries that use lithium include the lithium-ion polymer battery, lithium iron phosphate battery, and the nanowire battery.

Lubricating Greases

The third most common use of lithium is in greases. Lithium hydroxide is a strong base and, when heated with a fat, produces a soap made of lithium stearate. Lithium soap has the ability to thicken oils, and it is used to manufacture all-purpose, high-temperature lubricating greases.

Metallurgy

Lithium (e.g. as lithium carbonate) is used as an additive to continuous casting mould flux slags where it increases fluidity, a use which accounts for 5% of global lithium use (2011). Lithium compounds are also used as additives (fluxes) to foundry sand for iron casting to reduce veining.

Lithium (as lithium fluoride) is used as an additive to aluminium smelters (Hall–Héroult process), reducing melting temperature and increasing electrical resistance, a use which accounts for 3% of production (2011).

When used as a flux for welding or soldering, metallic lithium promotes the fusing of metals during the process and eliminates the forming of oxides by absorbing impurities. Alloys of the metal with aluminium, cadmium, copper and manganese are used to make high-performance aircraft parts.

Silicon Nano-welding

Lithium has been found effective in assisting the perfection of silicon nano-welds in electronic components for electric batteries and other devices.

Other Chemical and Industrial uses

Lithium use in flares and pyrotechnics is due to its rose-red flame.

Pyrotechnics

Lithium compounds are used as pyrotechnic colorants and oxidizers in red fireworks and flares.

Air Purification

Lithium chloride and lithium bromide are hygroscopic and are used as desiccants for gas streams. Lithium hydroxide and lithium peroxide are the salts most used in confined areas, such as aboard spacecraft and submarines, for carbon dioxide removal and air purification. Lithium hydroxide absorbs carbon dioxide from the air by forming lithium carbonate, and is preferred over other alkaline hydroxides for its low weight.

Lithium peroxide (Li_2O_2) in presence of moisture not only reacts with carbon dioxide to form lithium carbonate, but also releases oxygen. The reaction is as follows:

$$2\ Li_2O_2 + 2\ CO_2 \rightarrow 2\ Li_2CO_3 + O_2.$$

Some of the aforementioned compounds, as well as lithium perchlorate, are used in oxygen candles that supply submarines with oxygen. These can also include small amounts of boron, magnesium, aluminum, silicon, titanium, manganese, and iron.

Optics

Lithium fluoride, artificially grown as crystal, is clear and transparent and often used in specialist optics for IR, UV and VUV (vacuum UV) applications. It has one of the lowest refractive indexes and the farthest transmission range in the deep UV of most common materials. Finely divided lithium fluoride powder has been used for thermoluminescent radiation dosimetry (TLD): when a sample of such is exposed to radiation, it accumulates crystal defects which, when heated, resolve via a release of bluish light whose intensity is proportional to the absorbed dose, thus allowing this to be quantified. Lithium fluoride is sometimes used in focal lenses of telescopes.

The high non-linearity of lithium niobate also makes it useful in non-linear optics applications. It is used extensively in telecommunication products such as mobile phones and optical modulators, for such components as resonant crystals. Lithium applications are used in more than 60% of mobile phones.

Organic and Polymer Chemistry

Organolithium compounds are widely used in the production of polymer and fine-chemicals. In the polymer industry, which is the dominant consumer of these reagents, alkyl lithium compounds are catalysts/initiators. in anionic polymerization of unfunctionalizedolefins. For the production of fine chemicals, organolithium compounds function as strong bases and as reagents for the formation of carbon-carbon bonds. Organolithium compounds are prepared from lithium metal and alkyl halides.

Many other lithium compounds are used as reagents to prepare organic compounds. Some popular compounds include lithium aluminium hydride ($LiAlH_4$), lithium triethylborohydride, n-Butyllithium and tert-butyllithium are commonly used as extremely strong bases called superbase.

Military Applications

Metallic lithium and its complex hydrides, such as Li[AlH$_4$], are used as high-energy additives to rocket propellants. Lithium aluminum hydride can also be used by itself as a solid fuel.

The launch of a torpedo using lithium as fuel

The Mark 50 torpedo stored chemical energy propulsion system (SCEPS) uses a small tank of sulfur hexafluoride gas, which is sprayed over a block of solid lithium. The reaction generates heat, creating steam to propel the torpedo in a closed Rankine cycle.

Lithium hydride containing lithium-6 is used in thermonuclear weapons, where it encases the core of the bomb.

Nuclear

Lithium-6 is valued as a source material for tritium production and as a neutron absorber in nuclear fusion. Natural lithium contains about 7.5% lithium-6 from which large amounts of lithium-6 have been produced by isotope separation for use in nuclear weapons. Lithium-7 gained interest for use in nuclear reactorcoolants.

Lithium deuteride was used as fuel in the Castle Bravo nuclear device.

Lithium deuteride was the fusion fuel of choice in early versions of the hydrogen bomb. When bombarded by neutrons, both ^{6}Li and ^{7}Li produce tritium — this reaction, which was not fully understood when hydrogen bombs were first tested, was responsible for the runaway yield of the Castle Bravonuclear test. Tritium fuses with deuterium in a fusion reaction that is relatively easy to achieve. Although details remain secret, lithium-6 deuteride apparently still plays a role in modern nuclear weapons as a fusion material.

Lithium fluoride, when highly enriched in the lithium-7 isotope, forms the basic constituent of the fluoride salt mixture LiF-BeF$_2$ used in liquid fluoride nuclear reactors. Lithium fluoride is exceptionally chemically stable and LiF-BeF$_2$ mixtures have low melting points. In addition, ^{7}Li, Be, and F are among the few nuclides with low enough thermal neutron capture cross-sections not to poison the fission reactions inside a nuclear fission reactor.

In conceptualized (hypothetical) nuclear fusion power plants, lithium will be used to produce tritium in magnetically confined reactors using deuterium and tritium as the fuel. Naturally occurring tritium is extremely rare, and must be synthetically produced by surrounding the reacting plasma with a 'blanket' containing lithium where neutrons from the deuterium-tritium reaction in the plasma will fission the lithium to produce more tritium:

$$^6Li + n \rightarrow {}^4He + {}^3T.$$

Lithium is also used as a source for alpha particles, or helium nuclei. When ^7Li is bombarded by accelerated protons ^8Be is formed, which undergoes fission to form two alpha particles. This feat, called "splitting the atom" at the time, was the first fully man-made nuclear reaction. It was produced by Cockroft and Walton in 1932.

In 2013, the US Government Accountability Office said a shortage of lithium-7 critical to the operation of 65 out of 100 American nuclear reactors "places their ability to continue to provide electricity at some risk". The problem stems from the decline of US nuclear infrastructure. The equipment needed to separate lithium-6 from lithium-7 is mostly a cold war leftover. The US shut down most of this machinery in 1963, when it had a huge surplus of separated lithium, mostly consumed during the twentieth century. The report said it would take five years and $10 million to $12 million to reestablish the ability to separate lithium-6 from lithium-7.

Reactors that use lithium-7 heat water under high pressure and transfer heat through heat exchangers that are prone to corrosion. The reactors use lithium to counteract the corrosive effects of boric acid, which is added to the water to absorb excess neutrons.

Medicine

Lithium is useful in the treatment of bipolar disorder. Lithium salts may also be helpful for related diagnoses, such as schizoaffective disorder and cyclic major depression. The active part of these salts is the lithium ion Li^+. They may increase the risk of developing Ebstein's cardiac anomaly in infants born to women who take lithium during the first trimester of pregnancy.

Lithium has also been researched as a possible treatment for cluster headaches.

Precautions

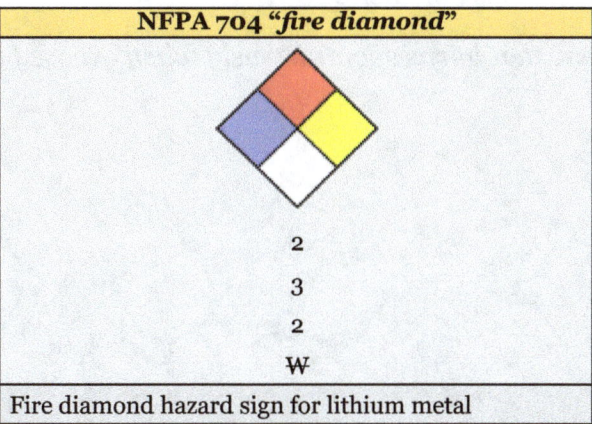

Fire diamond hazard sign for lithium metal

Lithium is corrosive and requires special handling to avoid skin contact. Breathing lithium dust or lithium compounds (which are often alkaline) initially irritate the nose and throat, while higher exposure can cause a buildup of fluid in the lungs, leading to pulmonary edema. The metal itself is a handling hazard because contact with moisture produces the caustic lithium hydroxide. Lithium is safely stored in non-reactive compounds such as naphtha.

Regulation

Some jurisdictions limit the sale of lithium batteries, which are the most readily available source of lithium for ordinary consumers. Lithium can be used to reduce pseudoephedrine and ephedrine to methamphetamine in the Birch reduction method, which employs solutions of alkali metals dissolved in anhydrous ammonia.

Carriage and shipment of some kinds of lithium batteries may be prohibited aboard certain types of transportation (particularly aircraft) because of the ability of most types of lithium batteries to fully discharge very rapidly when short-circuited, leading to overheating and possible explosion in a process called thermal runaway. Most consumer lithium batteries have built-in thermal over-load protection to prevent this type of incident, or are otherwise designed to limit short-circuit currents. Internal shorts from manufacturing defect or physical damage can lead to spontaneous thermal runaway.

Organometallic Compounds of Alkaline Metals

Organic compounds such as terminal alkynes which contain relatively acidic hydrogen atoms form salts with the alkali metals.

$$2\,Et-C \equiv C \cdot H + 2\,Na \rightarrow 2\,Na^{+} \lceil Me-C \equiv \sim C \rceil^{-} + H_{2}$$

$$Me-C \equiv C - H + K \lvert NH_{2} \rvert \rightarrow K^{+} \lceil Me-C \equiv C \rceil^{-} + NH_{3}$$

$$\text{(cyclopentadiene)} + NaH \rightarrow Na^{+}\ \text{(Cp}^{-}\text{)} + H_{2}$$

$$C_{5}H_{6} + Na \rightarrow NaCp + \tfrac{1}{2}\,H_{2}$$

NaCp is pyrophoric in air, but air-sensitivity can be lessened by complexing the Na^{+} with dme. In the solid state, [Na(dme)][Cp] is polymeric

Pyrophoric material: is one that burns spontaneously when exposed to air.

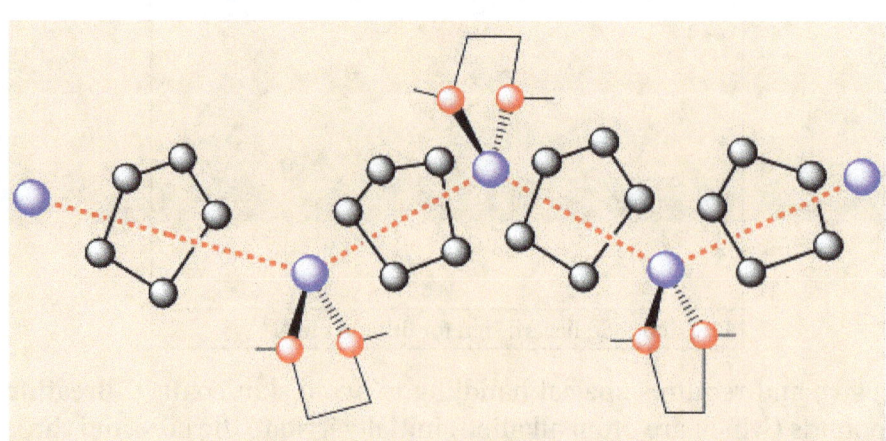

Transmetallation:

$$HgMe_2 + Na \rightarrow 2\,NaMe + Hg$$

Organolithium compounds:

$$^nBuCl + 2\,Li \xrightarrow{\substack{hydrocarbon \\ solvent}} {}^nBuLi + LiCl$$

Organolithium compounds are of particular importance among the group 1 organometallics.

Many of them are commercially available as solutions in hydrocarbon solvents.

Solvent choices for reactions involving organometallics of the alkali metals are critical. For example, ^nBuLi is decomposed by Et_2O to give ^nBuH, C_2H_4 and LiOEt.

Alkali metal organometallics are extremely reactive and must be handled in air- and moisture-free environments; NaMe, for example, burns explosively in air.

Lithium alkyls are polymeric both in solution and in the solid state.

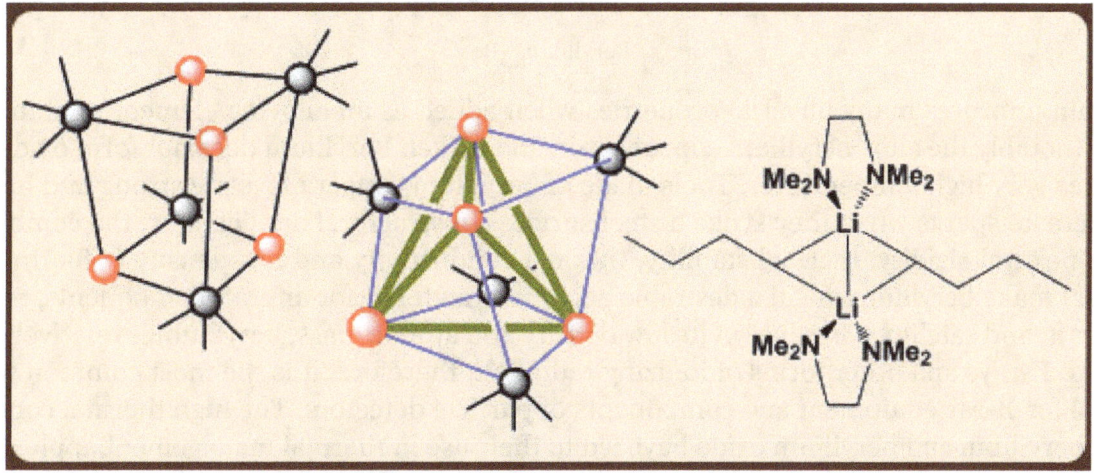

NMR is very useful in understanding the solution structures; ^6Li (I = 1), ^7Li (I = ½), ^{13}C (I = ½)

The structures of ($^tBuLi)_4$ and $(MeLi)_4$ are similar. ^nBuLi when mixed with TMEDA, gives a polymeric chain. TMEDA link cubane units together through the formation of Li-N bonds. Alkyllithium compounds are soluble in organic solvents whereas Na and K salts are insoluble, but are solubilized by the chelating ligand TMEDA. Addition of TMEDA may break down the aggregates of lithium alkyls to give lower nuclearity complexes. E.g. $[^nBuLi.TMEDA]_2$ However, detailed studies have revealed that the system is far from simple, and it is possible to isolate crystals of either $[^nBULi.TMEDA]_2$ or $[(^nBuLi)_4.TMEDA]_\infty$. In the case of $(MeLi)_4$, the addition of TMEDA does not lead to cluster breakdown, and the X-ray structure confirms the composition $(MeLi)_4.2TMEDA$, the presence of both tetramers and the amine molecules in the crystal lattice. Lithium alkyls and aryls are very useful reagents in organic synthesis and also in making corresponding carbon compounds of main group elements. Lithium alkyls are important catalysts in the synthetic rubber industry for the stereospecific polymerization of alkenes.

Beryllium

Beryllium is a chemical element with symbol Be and atomic number 4. It is a relatively rare element in the universe, usually occurring as a product of the spallation of larger atomic nuclei that have collided with cosmic rays. Within the cores of stars beryllium is depleted as it is fused and creates larger elements. It is a divalent element which occurs naturally only in combination with other elements in minerals. Notable gemstones which contain beryllium include beryl (aquamarine, emerald) and chrysoberyl. As a free element it is a steel-gray, strong, lightweight and brittle alkaline earth metal.

Beryllium, $_4$Be

Beryllium improves many physical properties when added as an alloying element to aluminium, copper (notably the alloy beryllium copper), iron and nickel. Beryllium does not form oxides until it reaches very high temperatures. Tools made of beryllium copperalloys are strong and hard and do not create sparks when they strike a steel surface. In structural applications, the combination of high flexural rigidity, thermal stability, thermal conductivity and low density (1.85 times that of water) make beryllium metal a desirable aerospace material for aircraft components, missiles, spacecraft, and satellites. Because of its low density and atomic mass, beryllium is relatively transparent to X-rays and other forms of ionizing radiation; therefore, it is the most common window material for X-ray equipment and components of particle detectors. The high thermal conductivities of beryllium and beryllium oxide have led to their use in thermal management applications.

The commercial use of beryllium requires the use of appropriate dust control equipment and industrial controls at all times because of the toxicity of inhaled beryllium-containing dusts that can cause a chronic life-threatening allergic disease in some people called berylliosis.

Characteristics

Physical Properties

Beryllium is a steel gray and hard metalloid (not a metal) that is brittle at room temperature and has a close-packed hexagonal crystal structure. It has exceptional stiffness (Young's modulus 287 GPa) and a reasonably high melting point. The modulus of elasticity of beryllium is approximately 50% greater than that of steel. The combination of this modulus and a relatively low density results in an unusually fast sound conduction speed in beryllium – about 12.9 km/s at ambient conditions. Other significant properties are high specific heat (1925 J·kg^{-1}·K^{-1}) and thermal conductivity (216 W·m^{-1}·K^{-1}), which make beryllium the metal with the best heat dissipation characteristics per unit weight. In combination with the relatively low coefficient of linear thermal expansion (11.4×10^{-6} K^{-1}), these characteristics result in a unique stability under conditions of thermal loading.

Nuclear Properties

Naturally occurring beryllium, save for slight contamination by cosmogenic radioisotopes, is essentially pure beryllium-9, which has a nuclear spin of 3/2. Beryllium has a large scattering cross section for high-energy neutrons, about 6 barns for energies above approximately 10 keV. Therefore, it works as a neutron reflector and neutron moderator, effectively slowing the neutrons to the thermal energy range of below 0.03 eV, where the total cross section is at least an order of magnitude lower – exact value strongly depends on the purity and size of the crystallites in the material.

The single primordial beryllium isotope ^{9}Be also undergoes a (n,2n) neutron reaction with neutron energies over about 1.9 MeV, to produce ^{8}Be, which almost immediately breaks into two alpha particles. Thus, for high-energy neutrons, beryllium is a neutron multiplier, releasing more neutrons than it absorbs. This nuclear reaction is:

$$^{9}_{4}Be + n \rightarrow 2(^{4}_{2}He) + 2n$$

Neutrons are liberated when beryllium nuclei are struck by energetic alpha particles producing the nuclear reaction

$$^{9}_{4}Be + ^{4}_{2}He \rightarrow ^{12}_{6}C + n \, , \quad \text{where } ^{4}_{2}He \text{ is an alpha particle and } ^{12}_{6}C \text{ is a carbon-12 nucleus.}$$

Beryllium also releases neutrons under bombardment by gamma rays. Thus, natural beryllium bombarded either by alphas or gammas from a suitable radioisotope is a key component of most radioisotope-powered nuclear reactionneutron sources for the laboratory production of free neutrons.

Small amounts of tritium are liberated when $^{9}_{4}Be$ nuclei absorb low energy neutrons in the three-step nuclear reaction

$$^{9}_{4}Be + n \rightarrow ^{4}_{2}He + ^{6}_{2}He, \quad ^{6}_{2}He \rightarrow ^{6}_{3}Li + \beta^{-}, \quad ^{6}_{3}Li + n \rightarrow ^{4}_{2}He + ^{3}_{1}H$$

Note that $^{6}_{2}He$ has a half life of only 0.8 seconds, β^{-} is an electron, and $^{6}_{3}Li$ has a high neutron absorption cross-section. Tritium is a radioisotope of concern in nuclear reactor waste streams.

As a metal, beryllium is transparent to most wavelengths of X-rays and gamma rays, making it useful for the output windows of X-ray tubes and other such apparatus.

Isotopes and Nucleosynthesis

Both stable and unstable isotopes of beryllium are created in stars, but the radioisotopes do not last long. It is believed that most of the stable beryllium in the universe was originally created in the interstellar medium when cosmic rays induced fission in heavier elements found in interstellar gas and dust. Primordial beryllium contains only one stable isotope, ^{9}Be, and therefore beryllium is a monoisotopic element.

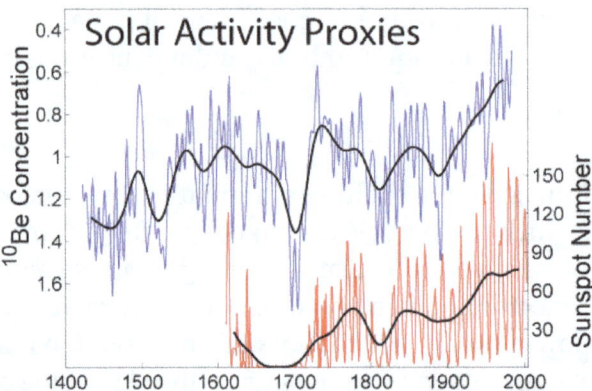

Plot showing variations in solar activity, including variation in sunspot number (red) and ^{10}Be concentration (blue). Note that the beryllium scale is inverted, so increases on this scale indicate lower ^{10}Be levels

Radioactive cosmogenic ^{10}Be is produced in the atmosphere of the Earth by the cosmic ray spallation of oxygen. ^{10}Be accumulates at the soil surface, where its relatively long half-life (1.36 million years) permits a long residence time before decaying to boron-10. Thus, ^{10}Be and its daughter products are used to examine natural soil erosion, soil formation and the development of lateritic soils, and as a proxy for measurement of the variations in solar activity and the age of ice cores. The production of ^{10}Be is inversely proportional to solar activity, because increased solar wind during periods of high solar activity decreases the flux of galactic cosmic rays that reach the Earth. Nuclear explosions also form ^{10}Be by the reaction of fast neutrons with ^{13}C in the carbon dioxide in air. This is one of the indicators of past activity at nuclear weapon test sites. The isotope ^{7}Be (half-life 53 days) is also cosmogenic, and shows an atmospheric abundance linked to sunspots, much like ^{10}Be.

^{8}Be has a very short half-life of about $7{\times}10^{-17}$ s that contributes to its significant cosmological role, as elements heavier than beryllium could not have been produced by nuclear fusion in the Big Bang. This is due to the lack of sufficient time during the Big Bang's nucleosynthesis phase to produce carbon by the fusion of ^{4}He nuclei and the very low concentrations of available beryllium-8. The British astronomer Sir Fred Hoyle first showed that the energy levels of ^{8}Be and ^{12}C allow carbon production by the so-called triple-alpha process in helium-fueled stars where more nucleosynthesis time is available. This process allows carbon to be produced in stars, but not in the Big Bang. Star-created carbon (the basis of carbon-based life) is thus a component in the elements in the gas and dust ejected by AGB stars and supernovae, as well as the creation of all other elements with atomic numbers larger than that of carbon.

The 2s electrons of beryllium may contribute to chemical bonding. Therefore, when ^{7}Be decays by L-electron capture, it does so by taking electrons from its atomic orbitals that may be participating in bonding. This makes its decay rate dependent to a measurable degree upon its chemical surroundings – a rare occurrence in nuclear decay.

The shortest-lived known isotope of beryllium is ^{13}Be which decays through neutron emission. It has a half-life of 2.7×10^{-21} s. ^{6}Be is also very short-lived with a half-life of 5.0×10^{-21} s. The exotic isotopes ^{11}Be and ^{14}Be are known to exhibit a nuclear halo. This phenomenon can be understood as the nuclei of ^{11}Be and ^{14}Be have, respectively, 1 and 4 neutrons orbiting substantially outside the classical Fermi 'waterdrop' model of the nucleus.

Occurrence

The Sun has a concentration of 0.1 parts per billion (ppb) of beryllium. Beryllium has a concentration of 2 to 6 parts per million (ppm) in the Earth's crust. It is most concentrated in the soils, 6 ppm. Trace amounts of ^{9}Be are found in the Earth's atmosphere. The concentration of beryllium in sea water is 0.2–0.6 parts per trillion. In stream water, however, beryllium is more abundant with a concentration of 0.1 ppb.

Beryllium ore with 1US¢ coin for scale

Emerald is a naturally occurring compound of beryllium.

Beryllium is found in over 100 minerals, but most are uncommon to rare. The more common beryllium containing minerals include: bertrandite ($Be_4Si_2O_7(OH)_2$), beryl ($Al_2Be_3Si_6O_{18}$), chrysoberyl (Al_2BeO_4) and phenakite (Be_2SiO_4). Precious forms of beryl are aquamarine, red beryl and emerald. The green color in gem-quality forms of beryl comes from varying amounts of chromium (about 2% for emerald).

The two main ores of beryllium, beryl and bertrandite, are found in Argentina, Brazil, India, Madagascar, Russia and the United States. Total world reserves of beryllium ore are greater than 400,000 tonnes.

Production

The extraction of beryllium from its compounds is a difficult process due to its high affinity for oxygen at elevated temperatures, and its ability to reduce water when its oxide film is removed. The United States, China and Kazakhstan are the only three countries involved in the industrial-scale extraction of beryllium.

Beryllium is most commonly extracted from the mineral beryl, which is either sintered using an extraction agent or melted into a soluble mixture. The sintering process involves mixing beryl with sodium fluorosilicate and soda at 770 °C (1,420 °F) to form sodium fluoroberyllate, aluminium oxide and silicon dioxide. Beryllium hydroxide is precipitated from a solution of sodium fluoroberyllate and sodium hydroxide in water. Extraction of beryllium using the melt method involves grinding beryl into a powder and heating it to 1,650 °C (3,000 °F). The melt is quickly cooled with water and then reheated 250 to 300 °C (482 to 572 °F) in concentrated sulfuric acid, mostly yielding beryllium sulfate and aluminium sulfate. Aqueous ammonia is then used to remove the aluminium and sulfur, leaving beryllium hydroxide.

Beryllium hydroxide created using either the sinter or melt method is then converted into beryllium fluoride or beryllium chloride. To form the fluoride, aqueous ammonium hydrogen fluoride is added to beryllium hydroxide to yield a precipitate of ammonium tetrafluoroberyllate, which is heated to 1,000 °C (1,830 °F) to form beryllium fluoride. Heating the fluoride to 900 °C (1,650 °F) with magnesium forms finely divided beryllium, and additional heating to 1,300 °C (2,370 °F) creates the compact metal. Heating beryllium hydroxide forms the oxide, which becomes beryllium chloride when combined with carbon and chlorine. Electrolysis of molten beryllium chloride is then used to obtain the metal.

Chemical Properties

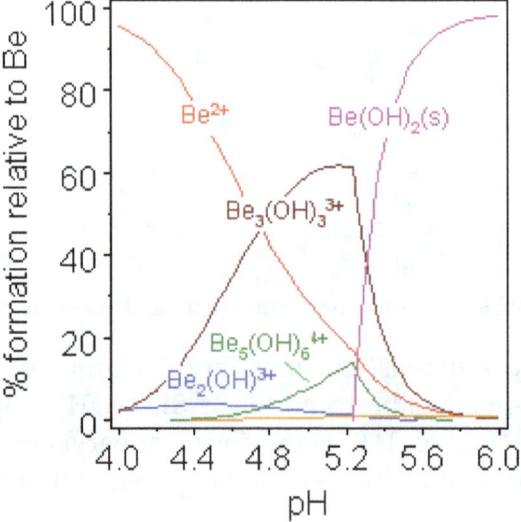

Beryllium hydrolysis as a function of pH Water molecules attached to Be are omitted

Beryllium's chemical behavior is largely a result of its small atomic and ionic radii. It thus has very high ionization potentials and strong polarization while bonded to other atoms, which is why all of its compounds are covalent. It is more chemically similar to aluminium than its close neighbors in the periodic table due to having a similar charge-to-radius ratio. An oxide layer forms around beryllium that prevents further reactions with air unless heated above 1000 °C. Once ignited, beryllium burns brilliantly forming a mixture of beryllium oxide and beryllium nitride. Beryllium dissolves readily in non-oxidizing acids, such as HCl and diluted H_2SO_4, but not in nitric acid or water as this forms the oxide. This behavior is similar to that of aluminium metal. Beryllium also dissolves in alkali solutions.

The beryllium atom has the electronic configuration [He] $2s^2$. The two valence electrons give beryllium a +2 oxidation state and thus the ability to form two covalent bonds; the only evidence of lower valence of beryllium is in the solubility of the metal in $BeCl_2$. Due to the octet rule, atoms tend to seek a valence of 8 in order to resemble a noble gas. Beryllium tries to achieve a coordination number of 4 because its two covalent bonds fill half of this octet. Tetracoordination allows beryllium compounds, such as the fluoride or chloride, to form polymers.

This characteristic is employed in analytical techniques using EDTA as a ligand. EDTA preferentially forms octahedral complexes – thus absorbing other cations such as Al^{3+} which might interfere – for example, in the solvent extraction of a complex formed between Be^{2+} and acetylacetone. Beryllium(II) readily forms complexes with strong donating ligands such as phosphine oxides and arsine oxides. There have been extensive studies of these complexes which show the stability of the O-Be bond.

Solutions of beryllium salts, e.g. beryllium sulfate and beryllium nitrate, are acidic because of hydrolysis of the $[Be(H_2O)_4]^{2+}$ ion.

$$[Be(H_2O)_4]^{2+} + H_2O \rightleftharpoons [Be(H_2O)_3(OH)]^+ + H_3O^+$$

Other products of hydrolysis include the trimeric ion $[Be_3(OH)_3(H_2O)_6]^{3+}$. Beryllium hydroxide, $Be(OH)_2$, is insoluble even in acidic solutions with pH less than 6, that is at biological pH. It is amphoteric and dissolves in strongly alkaline solutions.

Beryllium forms binary compounds with many non-metals. Anhydroushalides are known for F, Cl, Br and I. BeF_2 has a silica-like structure with corner-shared BeF_4 tetrahedra. $BeCl_2$ and $BeBr_2$ have chain structures with edge-shared tetrahedra. All beryllium halides have a linear monomeric molecular structure in the gas phase.

Beryllium difluoride, BeF_2, is different than the other difluorides. In general, beryllium has a tendency to bond covalently, much more so than the other alkaline earths and its fluoride is partially covalent (although still more ionic than its other halides). BeF_2 has many similarities to SiO_2 (quartz) a mostly covalently bonded network solid. BeF_2 has tetrahedrally coordinated metal and forms glasses (is difficult to crystallize). When crystalline, beryllium fluoride has the same room temperature crystal structure as quartz and shares many higher temperature structures also. Beryllium difluoride is very soluble in water, unlike the other alkaline earth difluorides. (Although they are strongly ionic, they do not dissolve because of the especially strong lattice energy of the fluorite structure.) However, BeF_2 has much lower electrical conductivity when in solution or when molten than would be expected if it were fully ionic.

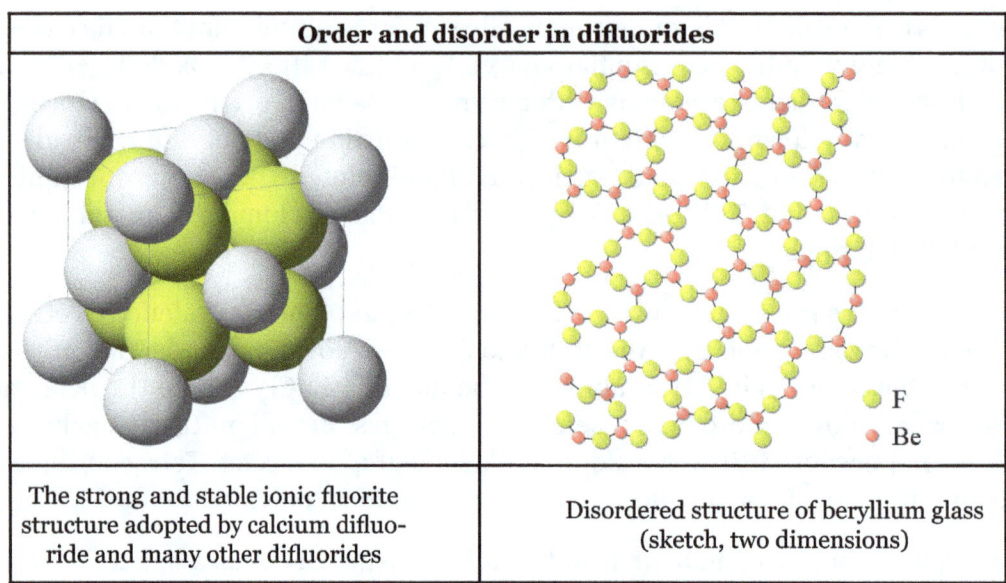

Order and disorder in difluorides

○ F
○ Be

| The strong and stable ionic fluorite structure adopted by calcium difluoride and many other difluorides | Disordered structure of beryllium glass (sketch, two dimensions) |

Beryllium oxide, BeO, is a white refractory solid, which has the wurtzite crystal structure and a thermal conductivity as high as in some metals. BeO is amphoteric. Salts of beryllium can be produced by treating $Be(OH)_2$ with acid. Beryllium sulfide, selenide and telluride are known, all having the zincblende structure.

Beryllium nitride, Be_3N_2 is a high-melting-point compound which is readily hydrolyzed. Beryllium azide, BeN_6 is known and beryllium phosphide, Be_3P_2 has a similar structure to Be_3N_2. Basic beryllium nitrate and basic beryllium acetate have similar tetrahedral structures with four beryllium atoms coordinated to a central oxide ion. A number of beryllium borides are known, such as Be_5B, Be_4B, Be_2B, BeB_2, BeB_6 and BeB_{12}. Beryllium carbide, Be_2C, is a refractory brick-red compound that reacts with water to give methane. No beryllium silicide has been identified.

History

The mineral beryl, which contains beryllium, has been used at least since the Ptolemaic dynasty of Egypt. In the first century CE, Roman naturalist Pliny the Elder mentioned in his encyclopedia *Natural History* that beryl and emerald ("smaragdus") were similar. The Papyrus Graecus Holmiensis, written in the third or fourth century CE, contains notes on how to prepare artificial emerald and beryl.

Louis-Nicolas Vauquelin discovered beryllium

Early analyses of emeralds and beryls by Martin Heinrich Klaproth, Torbern Olof Bergman, Franz Karl Achard, and Johann Jakob Bindheim always yielded similar elements, leading to the fallacious conclusion that both substances are aluminium silicates. Mineralogist René Just Haüy discovered that both crystals are geometrically identical, and he asked chemist Louis-Nicolas Vauquelin for a chemical analysis.

In a 1798 paper read before the Institut de France, Vauquelin reported that he found a new "earth" by dissolving aluminium hydroxide from emerald and beryl in an additional alkali. The editors of the journal *Annales de Chimie et de Physique* named the new earth "glucine" for the sweet taste of some of its compounds. Klaproth preferred the name "beryllina" due to the fact that yttria also formed sweet salts. The name "beryllium" was first used by Wöhler in 1828.

Friedrich Wöhler was one of the men who independently isolated beryllium

Friedrich Wöhler and Antoine Bussy independently isolated beryllium in 1828 by the chemical reaction of metallic potassium with beryllium chloride, as follows:

$$BeCl_2 + 2\ K \rightarrow 2\ KCl + Be$$

Using an alcohol lamp, Wöhler heated alternating layers of beryllium chloride and potassium in a wired-shut platinum crucible. The above reaction immediately took place and caused the crucible to become white hot. Upon cooling and washing the resulting gray-black powder he saw that it was made of fine particles with a dark metallic luster. The highly reactive potassium had been produced by the electrolysis of its compounds, a process discovered 21 years before. The chemical method using potassium yielded only small grains of beryllium from which no ingot of metal could be cast or hammered.

The direct electrolysis of a molten mixture of beryllium fluoride and sodium fluoride by Paul Lebeau in 1898 resulted in the first pure (99.5 to 99.8%) samples of beryllium. The first commercially successful process for producing beryllium was developed in 1932 by Alfred Stock and Hans Goldschmidt. Their process involves the electrolysis of a mixture of beryllium fluorides and barium, which causes molten beryllium to collect on a water-cooled iron cathode.

A sample of beryllium was bombarded with alpha rays from the decay of radium in a 1932 experiment by James Chadwick that uncovered the existence of the neutron. This same method is used in one class of radioisotope-based laboratory neutron sources that produce 30 neutrons for every million α particles.

Beryllium production saw a rapid increase during World War II, due to the rising demand for hard beryllium-copper alloys and phosphors for fluorescent lights. Most early fluorescent lamps used zinc orthosilicate with varying content of beryllium to emit greenish light. Small additions of magnesium tungstate improved the blue part of the spectrum to yield an acceptable white light. Halophosphate-based phosphors replaced beryllium-based phosphors after beryllium was found to be toxic.

Electrolysis of a mixture of beryllium fluoride and sodium fluoride was used to isolate beryllium during the 19th century. The metal's high melting point makes this process more energy-consuming than corresponding processes used for the alkali metals. Early in the 20th century, the production of beryllium by the thermal decomposition of beryllium iodide was investigated following the success of a similar process for the production of zirconium, but this process proved to be uneconomical for volume production.

Pure beryllium metal did not become readily available until 1957, even though it had been used as an alloying metal to harden and toughen copper much earlier. Beryllium could be produced by reducing beryllium compounds such as beryllium chloride with metallic potassium or sodium. Currently most beryllium is produced by reducing beryllium fluoride with purified magnesium. The price on the American market for vacuum-cast beryllium ingots was about $338 per pound ($745 per kilogram) in 2001.

Between 1998 and 2008, the world's production of beryllium had decreased from 343 to about 200 tonnes, of which 176 tonnes (88%) came from the United States.

Etymology

The original source is probably the Sanskrit word (*vaidurya*), which is of Dravidian origin and could be related to the name of the modern city of Belur. For about 160 years, beryllium was also known as glucinum or glucinium (with the accompanying chemical symbol "Gl", or "G"), the name coming from the Greek word for sweet: due to the sweet taste of beryllium salts.

Applications

Radiation Windows

Beryllium target which "converts" a proton beam into a neutron beam

A square beryllium foil mounted in a steel case to be used as a window between a vacuum chamber and an X-ray microscope. Beryllium is highly transparent to X-rays owing to its low atomic number.

Because of its low atomic number and very low absorption for X-rays, the oldest and still one of the most important applications of beryllium is in radiation windows for X-ray tubes. Extreme demands are placed on purity and cleanliness of beryllium to avoid artifacts in the X-ray images. Thin beryllium foils are used as radiation windows for X-ray detectors, and the extremely low absorption minimizes the heating effects caused by high intensity, low energy X-rays typical of synchrotron radiation. Vacuum-tight windows and beam-tubes for radiation experiments on synchrotrons are manufactured exclusively from beryllium. In scientific setups for various X-ray emission studies (e.g., energy-dispersive X-ray spectroscopy) the sample holder is usually made of beryllium because its emitted X-rays have much lower energies (~100 eV) than X-rays from most studied materials.

Low atomic number also makes beryllium relatively transparent to energetic particles. Therefore, it is used to build the beam pipe around the collision region in particle physics setups, such as all four main detector experiments at the Large Hadron Collider (ALICE, ATLAS, CMS, LHCb), the Tevatron and the SLAC. The low density of beryllium allows collision products to reach the surrounding detectors without significant interaction, its stiffness allows a powerful vacuum to be produced within the pipe to minimize interaction with gases, its thermal stability allows it to function correctly at temperatures of only a few degrees above absolute zero, and its diamagnetic nature keeps it from interfering with the complex multipole magnet systems used to steer and focus the particle beams.

Mechanical Applications

Because of its stiffness, light weight and dimensional stability over a wide temperature range, beryllium metal is used for lightweight structural components in the defense and aerospace industries in high-speed aircraft, guided missiles, spacecraft, and satellites. Several liquid-fuel rockets have used rocket nozzles made of pure beryllium. Beryllium powder was itself studied as a rocket fuel, but this use has never materialized. A small number of extreme high-end bicycle frames have been built with beryllium. From 1998 to 2000, the McLarenFormula One team used Mercedes-Benz engines with beryllium-aluminium-alloy pistons. The use of beryllium engine components was banned following a protest by Scuderia Ferrari.

Mixing about 2.0% beryllium into copper forms an alloy called beryllium copper that is six times stronger than copper alone. Beryllium alloys are used in many applications because of their combination of elasticity, high electrical conductivity and thermal conductivity, high strength and hardness, nonmagnetic properties, as well as good corrosion and fatigue resistance. These applications include non-sparking tools that are used near flammable gases (beryllium nickel), in springs and membranes (beryllium nickel and beryllium iron) used in surgical instruments and high temperature devices. As little as 50 parts per million of beryllium alloyed with liquid magnesium leads to a significant increase in oxidation resistance and decrease in flammability.

Beryllium Copper Adjustable Wrench

The high elastic stiffness of beryllium has led to its extensive use in precision instrumentation, e.g. in inertial guidance systems and in the support mechanisms for optical systems. Beryllium-copper alloys were also applied as a hardening agent in "Jason pistols", which were used to strip the paint from the hulls of ships.

Beryllium was also used for cantilevers in high performance phonograph cartridge styli, where its extreme stiffness and low density allowed for tracking weights to be reduced to 1 gram, yet still track high frequency passages with minimal distortion.

An earlier major application of beryllium was in brakes for military airplanes because of its hardness, high melting point, and exceptional ability to dissipate heat. Environmental considerations have led to substitution by other materials.

To reduce costs, beryllium can be alloyed with significant amounts of aluminium, resulting in the AlBeMet alloy (a trade name). This blend is cheaper than pure beryllium, while still retaining many desirable properties.

Mirrors

Beryllium mirrors are of particular interest. Large-area mirrors, frequently with a honeycomb support structure, are used, for example, in meteorological satellites where low weight and long-term dimensional stability are critical. Smaller beryllium mirrors are used in optical guidance systems and in fire-control systems, e.g. in the German-made Leopard 1 and Leopard 2 main battle tanks. In these systems, very rapid movement of the mirror is required which again dictates low mass and high rigidity. Usually the beryllium mirror is coated with hard electroless nickel plating which can be more easily polished to a finer optical finish than beryllium. In some applications, though, the beryllium blank is polished without any coating. This is particularly applicable to cryogenic operation where thermal expansion mismatch can cause the coating to buckle.

The James Webb Space Telescope will have 18 hexagonal beryllium sections for its mirrors. Because JWST will face a temperature of 33 K, the mirror is made of gold-plated beryllium, capable of handling extreme cold better than glass. Beryllium contracts and deforms less than glass – and remains more uniform – in such temperatures. For the same reason, the optics of the Spitzer Space Telescope are entirely built of beryllium metal.

Magnetic Applications

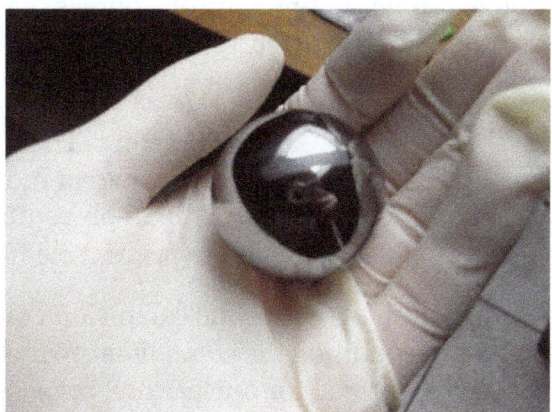

Sphere Beryllium B52 - Gyrocompass

Beryllium is non-magnetic. Therefore, tools fabricated out of beryllium-based materials are used by naval or military explosive ordnance disposal teams for work on or near naval mines, since these mines commonly have magnetic fuzes. They are also found in maintenance and construction materials near magnetic resonance imaging (MRI) machines because of the high magnetic fields generated. In the fields of radio communications and powerful (usually military) radars, hand tools made of beryllium are used to tune the highly magnetic klystrons, magnetrons, traveling wave tubes, etc., that are used for generating high levels of microwave power in the transmitters.

Nuclear Applications

Thin plates or foils of beryllium are sometimes used in nuclear weapon designs as the very outer layer of the plutonium pits in the primary stages of thermonuclear bombs, placed to surround the fissile material. These layers of beryllium are good "pushers" for the implosion of the plutonium-239, and they are also good neutron reflectors, just as they are in beryllium-moderated nuclear reactors.

Two CANDU fuel bundles: Each about 50 cm in length and 10 cm in diameter.
Notice the small appendages on the fuel clad surfaces

Beryllium is also commonly used in some neutron sources in laboratory devices in which relatively few neutrons are needed (rather than having to use a nuclear reactor, or a particle accelerator-powered neutron generator). For this purpose, a target of beryllium-9 is bombarded with energetic alpha particles from a radioisotope such as polonium-210, radium-226, plutonium-238, or americium-241. In the nuclear reaction that occurs, a beryllium nucleus is transmuted into carbon-12, and one free neutron is emitted, traveling in about the same direction as the alpha particle was heading. Such alpha decay driven beryllium neutron sources, named "urchin" neutron initiators, were used some in early atomic bombs. Neutron sources in which beryllium is bombarded with gamma rays from a gamma decay radioisotope, are also used to produce laboratory neutrons.

Beryllium is also used in fuel fabrication for CANDU reactors. The fuel elements have small appendages that are resistance brazed to the fuel cladding using an induction brazing process with Be as the braze filler material. Bearing pads are brazed on to prevent fuel bundle to pressure tube contact, and inter-element spacer pads are brazed on to prevent element to element contact.

Beryllium is also used at the Joint European Torusnuclear-fusion research laboratory, and it will be used in the more advanced ITER to condition the components which face the plasma. Beryllium has also been proposed as a cladding material for nuclear fuel rods, because of its good combination of mechanical, chemical, and nuclear properties. Beryllium fluoride is one of the constituent salts of the eutectic salt mixture FLiBe, which is used as a solvent, moderator and coolant in many hypothetical molten salt reactor designs, including the liquid fluoride thorium reactor (LFTR).

Acoustics

The low weight and high rigidity of beryllium make it useful as a material for high-frequency speaker drivers. Because beryllium is expensive (many times more than titanium), hard to shape due to its brittleness, and toxic if mishandled, beryllium tweeters are limited to high-end home,pro audio, and public address applications. Some high-fidelity products have been fraudulently claimed to be made of the material.

Electronic

Beryllium is a p-typedopant in III-V compound semiconductors. It is widely used in materials such as GaAs, AlGaAs, InGaAs and InAlAs grown by molecular beam epitaxy (MBE). Cross-rolled beryllium sheet is an excellent structural support for printed circuit boards in surface-mount technology. In critical electronic applications, beryllium is both a structural support and heat sink. The application also requires a coefficient of thermal expansion that is well matched to the alumina and polyimide-glasssubstrates. The beryllium-beryllium oxide composite "E-Materials" have been specially designed for these electronic applications and have the additional advantage that the thermal expansion coefficient can be tailored to match diverse substrate materials.

Beryllium oxide is useful for many applications that require the combined properties of an electrical insulator and an excellent heat conductor, with high strength and hardness, and a very high melting point. Beryllium oxide is frequently used as an insulator base plate in high-powertransistors in radio frequencytransmitters for telecommunications. Beryllium oxide is also being studied for use in increasing the thermal conductivity of uranium dioxidenuclear fuel pellets. Beryllium compounds were used in fluorescent lighting tubes, but this use was discontinued because of the disease berylliosis which developed in the workers who were making the tubes.

Healthcare

Beryllium is a component of several dental alloys.

Occupational Safety and Health

Beryllium is considered a health and safety issue for workers. Exposure to beryllium in the workplace can lead to a sensitization immune response and can over time develop chronic beryllium disease (CBD). The National Institute for Occupational Safety and Health (NIOSH) in the United States researches these effects in collaboration with a major manufacturer of beryllium products. The goal of this research is to prevent sensitization and CBD by developing a better understanding of the work processes and exposures that may present a potential risk for workers, and to develop effective interventions that will reduce the risk for adverse health effects. NIOSH also conducts genetic research on sensitization and CBD, independently of this collaboration. The NIOSH Manual of Analytical Methods contains methods for measuring occupational exposures to beryllium.

Precautions

Approximately 35 micrograms of beryllium is found in the average human body, an amount not considered harmful. Beryllium is chemically similar to magnesium and therefore can displace it from enzymes, which causes them to malfunction. Because Be^{2+} is a highly charged and small ion, it can easily get into many tissues and cells, where it specifically targets cell nuclei, inhibiting many enzymes, including those used for synthesizing DNA. Its toxicity is exacerbated by the fact that the body has no means to control beryllium levels, and once inside the body the beryllium cannot be removed. Chronic berylliosis is a pulmonary and systemicgranulomatous disease caused by inhalation of dust or fumes contaminated with beryllium; either large amounts over a short time or small amounts over a long time can lead to this ailment. Symptoms of the disease can take up to five years to develop; about a third of patients with it die and the survivors are left disabled. The International Agency for Research on Cancer (IARC) lists beryllium and beryllium compounds as Category 1 carcinogens. In the US, the Occupational Safety and Health Administration (OSHA) has designated a permissible exposure limit (PEL) in the workplace with a time-weighted average (TWA) 0.002 mg/m³ and a constant exposure limit of 0.005 mg/m³ over 30 minutes, with a maximum peak limit of 0.025 mg/m³. The National Institute for Occupational Safety and Health (NIOSH) has set a recommended exposure limit (REL) of constant 0.0005 mg/m³. The IDLH (immediately dangerous to life and health) value is 4 mg/m³.

The toxicity of finely divided beryllium (dust or powder, mainly encountered in industrial settings where beryllium is produced or machined) is very well-documented. Solid beryllium metal does not carry the same hazards as airborne inhaled dust, but any hazard associated with physical contact is poorly documented. Workers handling finished beryllium pieces are routinely advised to handle them with gloves, both as a precaution and because many if not most applications of beryllium cannot tolerate residue of skin contact such as fingerprints.

Acute beryllium disease in the form of chemical pneumonitis was first reported in Europe in 1933 and in the United States in 1943. A survey found that about 5% of workers in plants manufacturing fluorescent lamps in 1949 in the United States had beryllium-related lung diseases. Chronic berylliosis resembles sarcoidosis in many respects, and the differential diagnosis is often difficult. It killed some early workers in nuclear weapons design, such as Herbert L. Anderson.

Beryllium may be found in coal slag. When the slag is formulated into an abrasive agent for blasting paint and rust from hard surfaces, the beryllium can become airborne and become a source of exposure.

Early researchers tasted beryllium and its various compounds for sweetness in order to verify its presence. Modern diagnostic equipment no longer necessitates this highly risky procedure and no attempt should be made to ingest this highly toxic substance. Beryllium and its compounds should be handled with great care and special precautions must be taken when carrying out any activity which could result in the release of beryllium dust (lung cancer is a possible result of prolonged exposure to beryllium-laden dust). Although the use of beryllium compounds in fluorescent lighting tubes was discontinued in 1949, potential for exposure to beryllium exists in the nuclear and aerospace industries and in the refining of beryllium metal and melting of beryllium-containing alloys, the manufacturing of electronic devices, and the handling of other beryllium-containing material.

A successful test for beryllium in air and on surfaces has been recently developed and published as an international voluntary consensus standard ASTM D7202. The procedure uses dilute ammonium bifluoride for dissolution and fluorescence detection with beryllium bound to sulfonated hydroxybenzoquinoline, allowing up to 100 times more sensitive detection than the recommended limit for beryllium concentration in the workplace. Fluorescence increases with increasing beryllium concentration. The new procedure has been successfully tested on a variety of surfaces and is effective for the dissolution and ultratrace detection of refractory beryllium oxide and siliceous beryllium (ASTM D7458).

Magnesium

Magnesium, $_{12}$Mg

Magnesium is a chemical element with symbol Mg and atomic number 12. It is a shiny gray solid which bears a close physical resemblance to the other five elements in the second column (Group 2, or alkaline earth metals) of the periodic table: all Group 2 elements have the same electron configuration in the outer electron shell and a similar crystal structure.

Magnesium is the ninth most abundant element in the universe. It is produced in large, aging stars from the sequential addition of three helium nuclei to a carbon nucleus. When such stars explode as supernovas, much of the magnesium is expelled into the interstellar medium where it may recycle into new star systems. Magnesium is the eighth most abundant element in the Earth's crust and the fourth most common element in the Earth (after iron, oxygen and silicon), making up 13% of the planet's mass and a large fraction of the planet's mantle. It is the third most abundant element dissolved in seawater, after sodium and chlorine.

Magnesium occurs naturally only in combination with other elements, where it invariably has a +2 oxidation state. The free element (metal) can be produced artificially, and is highly reactive though in the atmosphere, it is soon coated in a thin layer of oxide that partly inhibits reactivity. The free metal burns with a characteristic brilliant-white light. The metal is now obtained mainly by electrolysis of magnesium salts obtained from brine, and is used primarily as a component in aluminium-magnesium alloys, sometimes called *magnalium* or *magnelium*. Magnesium is less dense than aluminium, and the alloy is prized for its combination of lightness and strength.

Magnesium is the eleventh most abundant element by mass in the human body and is essential to all cells and some 300 enzymes. Magnesium ions interact with polyphosphate compounds such as ATP, DNA, and RNA. Hundreds of enzymes require magnesium ions to function. Magnesium compounds are used medicinally as common laxatives, antacids (e.g., milk of magnesia), and to stabilize abnormal nerve excitation or blood vessel spasm in such conditions as eclampsia.

Characteristics

Physical Properties

Elemental magnesium is a gray-white lightweight metal, two-thirds the density of aluminium. It tarnishes slightly when exposed to air, although, unlike the other alkaline earth metals, an oxygen-free environment is unnecessary for storage because magnesium is protected by a thin layer of oxide that is fairly impermeable and difficult to remove. Magnesium has the lowest melting (923 K (1,202 °F)) and the lowest boiling point 1,363 K (1,994 °F) of all the alkaline earth metals.

Magnesium reacts with water at room temperature, though it reacts much more slowly than calcium, a similar group 2 metal. When submerged in water, hydrogen bubbles form slowly on the surface of the metal—though, if powdered, it reacts much more rapidly. The reaction occurs faster with higher temperatures. Magnesium's reversible reaction with water can be harnessed to store energy and run a magnesium-based engine.

Magnesium also reacts exothermically with most acids such as hydrochloric acid (HCl), producing the metal chloride and hydrogen gas, similar to the HCl reaction with aluminium, zinc, and many other metals.

Chemical Properties

Flammability

Magnesium is highly flammable, especially when powdered or shaved into thin strips, though it is difficult to ignite in mass or bulk. Flame temperatures of magnesium and magnesium alloys can reach 3,100 °C (3,370 K; 5,610 °F), although flame height above the burning metal is usually less than 300 mm (12 in). Once ignited, such fires are difficult to extinguish, with combustion continuing in nitrogen (forming magnesium nitride), carbon dioxide (forming magnesium oxide and carbon), and water (forming magnesium oxide and hydrogen). This property was used in incendiary weapons during the firebombing of cities in World War II, where the only practical civil defense was to smother a burning flare under dry sand to exclude atmosphere from the combustion.

Magnesium may also be used as an igniter for thermite, a mixture of aluminium and iron oxide powder that ignites only at a very high temperature.

Source of Light

When burning in air, magnesium produces a brilliant-white light that includes strong ultraviolet wavelengths. Magnesium powder (flash powder) was used for subject illumination in the early days of photography. Later, magnesium filament was used in electrically ignited single-use photography flashbulbs. Magnesium powder is used in fireworks and marine flares where a brilliant white light is required. It was also used for various theatrical effects, such as lightning, pistol flashes, and supernatural appearances.

Occurrence

Magnesium is the eighth-most-abundant element in the Earth's crust by mass and tied in seventh place with iron in molarity. It is found in large deposits of magnesite, dolomite, and other minerals, and in mineral waters, where magnesium ion is soluble.

Although magnesium is found in more than 60 minerals, only dolomite, magnesite, brucite, carnallite, talc, and olivine are of commercial importance.

The Mg cation is the second-most-abundant cation in seawater (about ⅛ the mass of sodium ions in a given sample), which makes seawater and sea salt attractive commercial sources for Mg. To extract the magnesium, calcium hydroxide is added to seawater to form magnesium hydroxideprecipitate.

$$MgCl_2 + Ca(OH)_2 \rightarrow Mg(OH)_2 + CaCl_2$$

Magnesium hydroxide (brucite) is insoluble in water and can be filtered out and reacted with hydrochloric acid to produced concentrated magnesium chloride.

$$Mg(OH)_2 + 2HCl \rightarrow MgCl_2 + 2H_2O$$

From magnesium chloride, electrolysis produces magnesium.

Forms

Alloy

As of 2013, magnesium alloy consumption was less than one million tons per year, compared with 50 million tons of aluminum alloys. Its use has been historically limited by its tendency to corrode, creep at high temperatures, and combust.

Corrosion

The presence of iron, nickel, copper, and cobalt strongly activates corrosion. Greater than a very small percentage, these metals precipitate as intermetallic compounds, and the precipitate locales function as active cathodic sites that reduce water, causing the loss of magnesium. Controlling the quantity of these metals improves corrosion resistance. Sufficient manganese overcomes the corrosive effects of iron. This requires precise control over composition, increasing costs. Adding a cathodic poison captures atomic hydrogen within the structure of a metal. This prevents the formation of free hydrogen gas, an essential factor of corrosive chemical processes. The addition of about one in three hundred parts arsenic reduces its corrosion rate in a salt solution by a factor of nearly ten.

High-temperature Creep and Flammability

Research showed that magnesium's tendency to creep at high-temperatures is eliminated by the adding scandium and gadolinium. Flammability is greatly reduced by a small amount of calcium in the alloy.

Compounds

Magnesium forms a variety of compounds important to industry and biology, including magnesium carbonate, magnesium chloride, magnesium citrate, magnesium hydroxide (milk of magnesia), magnesium oxide, magnesium sulfate, and magnesium sulfate heptahydrate (Epsom salts).

Isotopes

Magnesium has three stable isotopes: ^{24}Mg, ^{25}Mg and ^{26}Mg. All are present in significant amounts. About 79% of Mg is ^{24}Mg. The isotope ^{28}Mg is radioactive and in the 1950s to 1970s was produced by several nuclear power plants for use in scientific experiments. This isotope has a relatively short half-life (21 hours) and its use was limited by shipping times.

The isomer ^{26}Mg has found application in isotopicgeology, similar to that of aluminium. ^{26}Mg is a radiogenic daughter product of ^{26}Al, which has a half-life of 717,000 years. Excessive quantities of stable ^{26}Mg have been observed in the Ca-Al-rich inclusions of some meteorites. This anomalous abundance is attributed to the decay of its parent ^{26}Al in the inclusions, and researchers conclude that such meteorites were formed in the solar nebula before the ^{26}Al had decayed. These are among the oldest objects in the solar system and contain preserved information about its early history.

It is conventional to plot $^{26}Mg/^{24}Mg$ against an Al/Mg ratio. In an isochron dating plot, the Al/Mg ratio plotted is $^{27}Al/^{24}Mg$. The slope of the isochron has no age significance, but indicates the initial $^{26}Al/^{27}Al$ ratio in the sample at the time when the systems were separated from a common reservoir.

Production

China is the dominant supplier of magnesium, with approximately 80% of the world market share. China is almost completely reliant on the silicothermicPidgeon process (the reduction of the oxide at high temperatures with silicon, often provided by a ferrosilicon alloy in which the iron is but a spectator in the reactions) to obtain the metal. The process can also be carried out with carbon at approx 2300 °C:

Country	2011 production (tonnes)
China	661,000
U.S.	63,500
Russia	37,000
Israel	30,000
Kazakhstan	21,000
Brazil	16,000
Ukraine	2,000
Serbia	1,500
Total	832,000

Magnesium sheets and ingots

$$2MgO_{(s)} + Si_{(s)} + 2CaO_{(s)} \rightarrow 2Mg_{(g)} + Ca_2SiO_{4(s)}$$

$$MgO_{(s)} + C_{(s)} \rightarrow Mg_{(g)} + CO_{(g)}$$

In the United States, magnesium is obtained principally with the Dow process, by electrolysis of fused magnesium chloride from brine and sea water. A saline solution containing Mg2+ ions is first treated with lime (calcium oxide) and the precipitated magnesium hydroxide is collected:

$$Mg^{2+}_{(aq)} + CaO_{(s)} + H_2O \rightarrow Ca^{2+}_{(aq)} + Mg(OH)_{2(s)}$$

The hydroxide is then converted to a partial hydrate of magnesium chloride by treating the hydroxide with hydrochloric acid and heating of the product:

$$Mg(OH)_{2(s)} + 2\ HCl \rightarrow MgCl_{2(aq)} + 2H_2O_{(l)}$$

The salt is then electrolyzed in the molten state. At the cathode, the Mg2+ ion is reduced by two electrons to magnesium metal:

$$Mg^{2+} + 2e^- \rightarrow Mg$$

At the anode, each pair of Cl⁻ions is oxidized to chlorine gas, releasing two electrons to complete the circuit:

$$2Cl^- \rightarrow Cl_2(g) + 2e^-$$

A new process, solid oxide membrane technology, involves the electrolytic reduction of MgO. At the cathode, Mg^{2+}ion is reduced by two electrons to magnesium metal. The electrolyte is Yttria-stabilized zirconia (YSZ). The anode is a liquid metal. At the YSZ/liquid metal anode $O^{2-}$is oxidized. A layer of graphite borders the liquid metal anode, and at this interface carbon and oxygen react to form carbon monoxide. When silver is used as the liquid metal anode, there is no reductant carbon or hydrogen needed, and only oxygen gas is evolved at the anode. It has been reported that this method provides a 40% reduction in cost per pound over the electrolytic reduction method. This method is more environmentally sound than others because there is much less carbon dioxide emitted.

The United States has traditionally been the major world supplier of this metal, supplying 45% of world production even as recently as 1995. Today, the US market share is at 7%, with a single domestic producer left, US Magnesium, a Renco Group company in Utah born from now-defunct Magcorp.

History

The name magnesium originates from the Greek word for a district in Thessaly called Magnesia. It is related to magnetite and manganese, which also originated from this area, and required differentiation as separate substances.

In 1618, a farmer at Epsom in England attempted to give his cows water from a well there. The cows refused to drink because of the water's bitter taste, but the farmer noticed that the water seemed to heal scratches and rashes. The substance became known as Epsom salts and its fame spread. It was eventually recognized as hydrated magnesium sulfate, $MgSO_4 \cdot 7H_2O$.

The metal itself was first isolated by Sir Humphry Davy in England in 1808. He used electrolysis on a mixture of magnesia and mercuric oxide.Antoine Bussy prepared it in coherent form in 1831. Davy's first suggestion for a name was magnium, but the name magnesium is now used.

Uses as a Metal

Magnesium is the third-most-commonly-used structural metal, following iron and aluminium. It is called the lightest useful metal by The Periodic Table of Videos.

The main applications of magnesium are, in order: aluminium alloys, die-casting (alloyed with zinc), removing sulfur in the production of iron and steel, and the production of titanium in the Kroll process.

Magnesium is used in super-strong, lightweight materials and alloys. For example, when infused with silicon carbide nanoparticles, it has extremely high specific strength.

An unusual application of magnesium as an illumination source while wakeskating in 1931

Historically, magnesium was one of the main aerospace construction metals and was used for German military aircraft as early as World War I and extensively for German aircraft in World War II.

The Germans coined the name "Elektron" for magnesium alloy, a term is still used today. In the commercial aerospace industry, magnesium was generally restricted to engine-related components, due fire and corrosion hazards. Currently, magnesium alloy use in aerospace is increasing, driven by the importance of fuel economy. Development and testing of new magnesium alloys continues, notably Elektron 21, which (in test) has proved suitable for aerospace engine, internal, and airframe components. The European Community runs three R&D magnesium projects in the Aerospace priority of Six Framework Program.

In the form of thin ribbons, magnesium is used to purify solvents; for example, preparing super-dry ethanol.

Aircraft

- Wright Aeronautical used a magnesium crankcase in the WWII-era Wright Duplex Cyclone aviation engine. This presented a serious problem for the earliest models of the Boeing B-29 heavy bomber when an in-flight engine fire ignited the engine crankcase. The resulting combustion was as hot as 5,600 °F (3,100 °C) and could sever the wing spar from the fuselage.

Automotive

Mg alloy motorcycle engine blocks

- Mercedes-Benz used the alloy Elektron in the body of an early model Mercedes-Benz 300 SLR; these cars ran (with successes) at Le Mans, the Mille Miglia, and other world-class race events in 1955.

- Porsche used magnesium alloy frames in the 917/053 that won Le Mans in 1971, and continues to use magnesium alloys for its engine blocks due to the weight advantage.

- Volkswagen Group has used magnesium in its engine components for many years.

- Mitsubishi Motors uses magnesium for its paddle shifters.

- BMW used magnesium alloy blocks in their N52 engine, including an aluminium alloy insert for the cylinder walls and cooling jackets surrounded by a high-temperature magnesium alloy AJ62A. The engine was used worldwide between 2005 and 2011 in various 1, 3, 5, 6, and 7 series models; as well as the Z4, X1, X3, and X5.

- Chevrolet used the magnesium alloy AE44 in the 2006 Corvette Z06.

Both AJ62A and AE44 are recent developments in high-temperature low-creep magnesium alloys. The general strategy for such alloys is to form intermetallic precipitates at the grain boundaries, for example by adding mischmetal or calcium. New alloy development and lower costs that make magnesium competitive with aluminium will increase the number of automotive applications.

Electronics

Because of low weight and good mechanical and electrical properties, magnesium is widely used for manufacturing of mobile phones, laptop and tablet computers, cameras, and other electronic components.

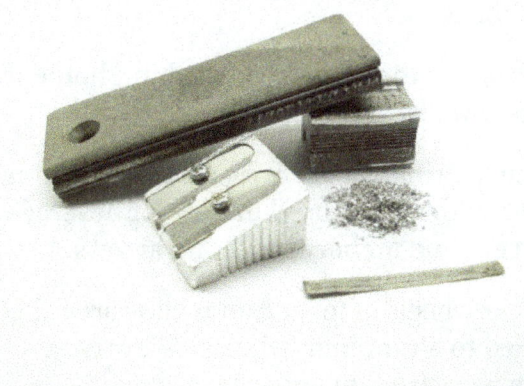

Products made of magnesium: firestarter and shavings, sharpener, magnesium ribbon

Other

Magnesium, being readily available and relatively nontoxic, has a variety of uses:

- Magnesium is flammable, burning at a temperature of approximately 3,100 °C (3,370 K; 5,610 °F), and the autoignition temperature of magnesium ribbon is approximately 473 °C (746 K; 883 °F). It produces intense, bright, white light when it burns. Magnesium's high combustion temperature makes it a useful tool for starting emergency fires. Other uses include flash photography, flares, pyrotechnics, and fireworks sparklers. Magnesium is also often used to ignite thermite or other materials that require a high ignition temperature.

Magnesium firestarter (in left hand), used with a pocket knife
and flint to create sparks that ignite the shavings

- In the form of turnings or ribbons, to prepare Grignard reagents, which are useful in organic synthesis.

- As an additive agent in conventional propellants and the production of nodular graphite in cast iron.

- As a reducing agent to separate uranium and other metals from their salts.

- As a sacrificial (galvanic) anode to protect boats, underground tanks, pipelines, buried structures, and water heaters.

- Alloyed with zinc to produce the zinc sheet used in photoengraving plates in the printing industry, dry-cell battery walls, and roofing.

- As a metal, this element's principal use is as an alloying additive to aluminium with these aluminium-magnesium alloys being used mainly for beverage cans, sports equipment such as golf clubs, fishing reels, and archery bows and arrows.

- Specialty, high-grade car wheels of magnesium alloy are called "mag wheels", although the term is often misapplied to aluminium wheels. Many car and aircraft manufacturers have made engine and body parts from magnesium.

- Magnesium batteries have been commercialized as primary batteries, and are an active topic of research for rechargeable secondary batteries.

Safety Precautions

Magnesium metal and its alloys can be explosive hazards; they are highly flammable in their pure form when molten or in powder or ribbon form. Burning or molten magnesium reacts violently with water. When working with powdered magnesium, safety glasses with eye protection and UV filters (such as welders use) are employed because burning magnesium produces ultraviolet light that can permanently damage the retina of a human eye.

The combusting magnesium-bodied Honda RA302 at the 1968 French Grand Prix, after the crash that killed driver Jo Schlesser.

Magnesium is capable of reducing water and releasing highly flammable hydrogen gas:

$$Mg\ (s)\ +\ 2H_2O(l)\ \rightarrow Mg(OH)_2(s)\ +H_2(g)$$

Therefore, water cannot extinguish magnesium fires. The hydrogen gas produced intensifies the fire. Dry sand is an effective smothering agent, but only on relatively level and flat surfaces.

Magnesium reacts with carbon dioxide exothermically to form magnesium oxide and carbon:

$$2\ Mg\ +CO_2 \rightarrow\ 2\ MgO\ +\ C\ (s)$$

Hence, carbon dioxide fuels rather than extinguishes magnesium fires.

Burning magnesium can be quenched by using a Class D dry chemical fire extinguisher, or by covering the fire with sand or magnesium foundry flux to remove its air source.

Useful Compounds

Magnesium compounds, primarily magnesium oxide (MgO), are used as a refractory material in furnace linings for producing iron, steel, nonferrous metals, glass, and cement. Magnesium oxide and other magnesium compounds are also used in the agricultural, chemical, and construction industries. Magnesium oxide from calcination is used as an electrical insulator in fire-resistant cables.

Magnesium reacted with an alkyl halide gives a Grignard reagent, which is a very useful tool for preparing alcohols.

Magnesium salts are included in various foods, fertilizers (magnesium is a component of chlorophyll), and microbe culture media.

Magnesium sulfite is used in the manufacture of paper (sulfite process).

Magnesium phosphate is used to fireproof wood used in construction.

Magnesium hexafluorosilicate is used for moth-proofing textiles.

Mechanism of Action

The important interaction between phosphate and magnesium ions makes magnesium essential to the basic nucleic acid chemistry of all cells of all known living organisms. More than 300 enzymes require magnesium ions for their catalytic action, including all enzymes using or synthesizing ATP and those that use other nucleotides to synthesize DNA and RNA. The ATP molecule is normally found in a chelate with a magnesium ion.

Dietary sources, recommended intake, and supplementation:

Examples of food sources of magnesium

Spices, nuts, cereals, cocoa and vegetables are rich sources of magnesium. Green leafy vegetables such as spinach are also rich in magnesium.

In the UK, the recommended daily values for magnesium is 300 mg for men and 270 mg for women. In the U.S. the Recommended Dietary Allowances (RDAs) are 400 mg for men ages 19–30 and 420 mg for older; for women 310 mg for ages 19–30 and 320 mg for older.

Numerous pharmaceutical preparations of magnesium and dietary supplements are available. In two human trials magnesium oxide, one of the most common forms in magnesium dietary supplements because of its high magnesium content per weight, was less bioavailable than magnesium citrate, chloride, lactate or aspartate.

Metabolism

An adult has 22–26 grams of magnesium, with 60% in the skeleton, 39% intracellular (20% in skeletal muscle), and 1% extracellular. Serum levels are typically 0.7–1.0 mmol/L or 1.8–2.4 mEq/L. Serum magnesium levels may be normal even when intracellular magnesium is deficient. The mechanisms for maintaining the magnesium level in the serum are varying gastrointestinal absorption and renal excretion. Intracellular magnesium is correlated with intracellular potassium. Increased magnesium lowers calcium and can either prevent hypercalcemia or cause hypocalcemia depending on the initial level. Both low and high protein intake conditions inhibit magnesium absorption, as does the amount of phosphate, phytate, and fat in the gut. Unabsorbed dietary magnesium is excreted in feces; absorbed magnesium is excreted in urine and sweat.

Detection in Serum and Plasma

Magnesium status may be assessed by measuring serum and erythrocyte magnesium concentrations coupled with urinary and fecal magnesium content, but intravenous magnesium loading tests are more accurate and practical. A retention of 20% or more of the injected amount indicates deficiency. No biomarker has been established for magnesium.

Magnesium concentrations in plasma or serum may be monitored for efficacy and safety in those receiving the drug therapeutically, to confirm the diagnosis in potential poisoning victims, or to assist in the forensic investigation in a case of fatal overdose. The newborn children of mothers who received parenteral magnesium sulfate during labor may exhibit toxicity with normal serum magnesium levels.

Deficiency

Magnesium deficiency (hypomagnesemia) is common: it is found in 2.5–15% of the general population. The primary cause of deficiency is decreased dietary intake: only 32% of people in the United States meet the recommended daily allowance. Other causes are increased renal or gastrointestinal loss, an increased intracellular shift, and proton-pump inhibitor antacid therapy. Most are asymptomatic, but symptoms referable to neuromuscular, cardiovascular, and metabolic dysfunction may occur. Alcoholism is often associated with magnesium deficiency. Chronically low serum magnesium levels are associated with metabolic syndrome, diabetes mellitus type 2, fasciculation, and hypertension.

Therapy

- Intravenous magnesium is recommended by the ACC/AHA/ESC 2006 Guidelines for Management of Patients With Ventricular Arrhythmias and the Prevention of Sudden Cardiac Death for patients with ventricular arrhythmia associated with torsades de pointes who present with long QT syndrome; and for the treatment of patients with digoxin induced arrhythmias.

- Magnesium is the drug of choice in the management of pre-eclampsia and eclampsia.

- Hypomagnesemia, including that caused by alcoholism, is reversible by oral or parenteral magnesium administration depending on the degree of deficiency.

- There is limited evidence that magnesium supplementation may play a role in the prevention and treatment of migraine.

- Oral magnesium may be therapeutic for restless leg syndrome.

Sorted by type of magnesium salt, other therapeutic applications include:

- Magnesium sulfate, as the heptahydrate called Epsom salts, is used as bath salts, a laxative, and a highly soluble fertilizer.

- Magnesium hydroxide, suspended in water, is used in milk of magnesiaantacids and laxatives.

- Magnesium chloride, oxide, gluconate, malate, orotate, glycinate, ascorbate and citrate are all used as oral magnesium supplements.

- Magnesium borate, magnesium salicylate, and magnesium sulfate are used as antiseptics.

- Magnesium bromide is used as a mild sedative (this action is due to the bromide, not the magnesium).

- Magnesium stearate is a slightly flammable white powder with lubricating properties. In pharmaceutical technology, it is used in pharmacological manufacture to prevent tablets from sticking to the equipment while compressing the ingredients into tablet form.

- Magnesium carbonate powder is used by athletes such as gymnasts, weightlifters, and climbers to eliminate palm sweat, prevent sticking, and improve the grip on gymnastic apparatus, lifting bars, and climbing rocks.

- Magnesium L–threonate has been studied as a possible treatment of mild cognitive impairment.

Overdose

Overdose from dietary sources alone is unlikely because excess magnesium in the blood is promptly filtered by the kidneys, and overdose is more likely in the presence of impaired renal function. In spite of this, megadose therapy has caused death in a young child, and severe hypermagnesemia in a woman and a young girl who had healthy kidneys. The most common symptoms of overdose are nausea, vomiting, and diarrhea; other symptoms include hypotension, confusion, slowed heart and respiratory rate, deficiencies of other minerals, coma, cardiac arrhythmia, and death from cardiac arrest.

Function in Plants

Plants require magnesium to synthesize chlorophyll, essential for photosynthesis. Magnesium in the center of the porphyrin ring in chlorophyll functions in a manner similar to the iron in the center of the porphyrin ring in heme. Magnesium deficiency in plants causes late-season yellowing between leaf veins, especially in older leaves, and can be corrected by applying to the soil either Epsom salts (which is rapidly leached), or crushed dolomiticlimestone.

Organometallic compounds of alkaline earth metals (beryllium and magnesium).

Beryllium

$$HgMe_2 + Be \rightarrow Me_2Be + Hg \quad (at\ 383\ K)$$

$$2PhLi + BeCl_2 \rightarrow Ph_2Be + 2LiCl \quad (in\ diethyl\ ether)$$

In vapor phase, Me2Be is monomeric with a linear C—Be—C (Be-C = 170 pm).

The solid state structure is polymeric and resembles that of $BeCl_2$.

$$2NaCp + BeCl_2 \rightarrow Cp_2Be + 2NaCl$$

The X-ray diffraction at 128 K suggested [(η^1-Cp)(η^5-Cp)Be].

However, ^{1}H NMR spectrum shows that all protons environments are equivalent even at 163 K.

Also, solid state structure shows the Be atom is disordered over two equivalent sites and NMR data can be interpreted in terms of fluxional process in which the Be atom moves between these two sites.

However, Cp*$_2$Be possesses a sandwich structure with both the rings are coplanar.

Magnesium

Alkyl and aryl magnesium halides (Grignard reagents, RMgX) are extremely well-known on account of their uses in synthetic chemistry.

$$Mg + RX \rightarrow RMgX \ (in\ diethyl\ ether)$$

Transmetallation is useful means of preparing pure Grignard reagents

$$Mg + RHgBr \rightarrow Hg + RMgBr$$

$$Mg + R_2Hg \rightarrow Hg + R_2Mg$$

Two-coordination at Mg in R_2Mg is observed only when the R groups are sufficiently bulky, e.g. Mg{C(SiMe$_3$)$_3$}$_2$. RMgX are generally solvated and Mg centre is typically tetrahedral. e.g. EtMgBr.2Et$_2$O; PhMgBr.2Et$_2$O.

Cp$_2$Mg has a staggered sandwich structure.

Solutions of Grignard reagent may contain several species, e.g. RMgX, R$_2$Mg, MgX$_2$, RMg(μ-X)$_2$MgR, which are further complicated by solvation. The position of equilibrium between these species is markedly dependent on concentration, temperature and solvent; strongly donating solvents favour monomeric species in which they coordinate to the metal centre.

$$2RMgX \quad \rightleftharpoons \quad R_2Mg + MgX_2$$

Treatment with dioxane results in the precipitation of $MgCl_2$(dioxane) leaving behind pure R_2Mg in the solution.

Structure and Bonding

The slight differences that arise between organometallic compounds and binary hydrogen compounds are mainly due to the tendency of alkyl groups to avoid ionic bonding.

The molecular structures of $AlMe_3$ and MeLi differ from AlH_3 and LiH. Even the more ionic MeK crystallizes in the nickel-arsenide structure rather than the rock-salt structure adopted by KCl.

Nickel-arsenide structure is typical of soft-cation, soft-anion combinations.

Electron deficient compounds such as $AlMe_3$ contain 3c-2e bonds analogous to the B—H—B bridges in diborane.

The Nickel-Arsenide, NiAs, Structure

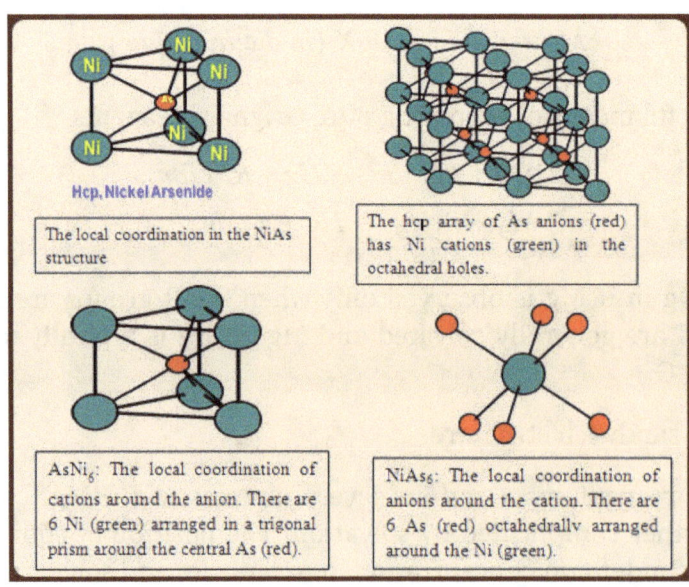

Hcp, Nickel Arsenide

The local coordination in the NiAs structure

The hcp array of As anions (red) has Ni cations (green) in the octahedral holes.

$AsNi_6$: The local coordination of cations around the anion. There are 6 Ni (green) arranged in a trigonal prism around the central As (red).

$NiAs_6$: The local coordination of anions around the cation. There are 6 As (red) octahedrally arranged around the Ni (green).

MeLi in nonpolar solvents consists of tetrahedron of Li atoms with each face bridged by a methyl group. Similar to Al_2Me_6, the bonding in MeLi consists of a set of localized molecular orbitals. The symmetric combination of three Li 2s orbitals on each face of the Li_4 tetrahedron and one sp^3 hybrid orbital from CH_3 gives an orbital that can accommodate a pair of electron to form a 4c-2e bond.

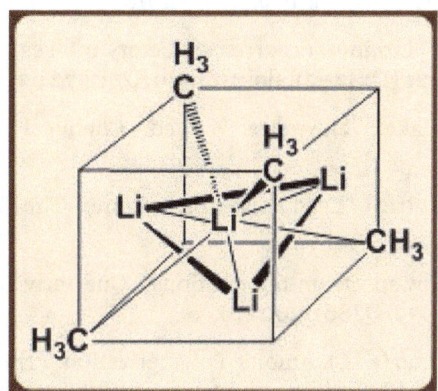

The lower energy of the C orbital compared with the Li orbitals indicates that the bonding pair of electrons will be associated primarily with the CH_3 group, thus supporting the carbanionic character of the molecule. Some analysis has indicated that about 90% ionic character for the Li-CH_3 interaction.

The interaction between an sp3 orbital from a methyl group and the three 2s orbitals of the Li atoms in a triangular face of Li4(CH3)4 to form a totally symmetric 4c,2e bonding orbital. The next higher orbital is non-bonding and the uppermost is antibonding.

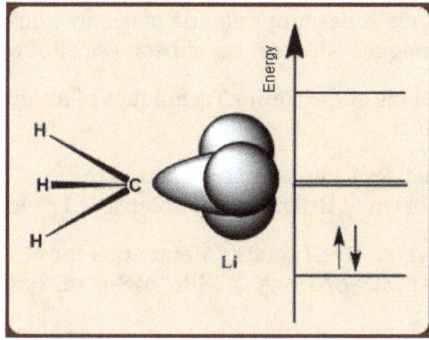

Me_2Be and Me_2Mg exist in a polymeric structure with two 3c,2e-bonding CH_3 bridges between each metal atom.

References

- Weeks, Mary Elvira (1932). "The discovery of the elements. IX. Three alkali metals: Potassium, sodium, and lithium". Journal of Chemical Education. 9 (6): 1035. Bibcode:1932JChEd...9.1035W. doi:10.1021/ed009p1035

- Lide, David R. (2003-06-19). CRC Handbook of Chemistry and Physics, 84th Edition. CRC Handbook. CRC Press. 14: Abundance of Elements in the Earth's Crust and in the Sea. ISBN 978-0-8493-0484-2

- Lavelle, Laurence. "Lanthanum (La) and Actinium (Ac) Should Remain in the d-Block" (PDF). lavelle.chem. ucla.edu. Retrieved 9 November 2014

- Banks, Alton (1990). "Sodium". Journal of Chemical Education. 67 (12): 1046. Bibcode:1990JChEd..67.1046B. doi:10.1021/ed067p1046

- B. Pearson (ed.). Speciality Chemicals: Innovations in industrial synthesis and applications (illustrated ed.). Springer Science & Business Media, 1991. p. 260. ISBN 1-85166-646-X

- Wilford, John Noble (17 March 1983). "ROMAN EMPIRE'S FALL IS LINKED WITH GOUT AND LEAD POISONING". The New York Times. Retrieved 19 January 2016

- Schrauzer, Gerhard N. (2002). "Lithium: occurrence, dietary intakes, nutritional essentiality". Journal of the American College of Nutrition. 21 (1): 14–21. doi:10.1080/07315724.2002.10719188. PMID 11838882

- Newton, David E. (1999). Baker, Lawrence W., ed. Chemical Elements. ISBN 978-0-7876-2847-5. OCLC 39778687

- Killgrove, Kristina (20 January 2012). "Lead Poisoning in Rome - The Skeletal Evidence". Powered by Osteons. Retrieved 19 January 2016

- "XXIV.—On chemical analysis by spectrum-observations". Quarterly Journal of the Chemical Society of London. 13 (3): 270. 1861. doi:10.1039/QJ8611300270

- Nikos Hadjichristidis; Akira Hirao (eds.). Anionic Polymerization: Principles, Practice, Strength, Consequences and Applications (illustrated ed.). Springer, 2015. p. 349. ISBN 4-431-54186-1

- Sumner, Thomas (21 April 2014). "Did Lead Poisoning Bring Down Ancient Rome?". Science Magazine. Retrieved 19 January 2016

- Asplund, M.; et al. (2006). "Lithium Isotopic Abundances in Metal-poor Halo Stars". The Astrophysical Journal. 644: 229–259. arXiv:astro-ph/0510636. Bibcode:2006ApJ...644..229A. doi:10.1086/503538

- Xu Hou (ed.). Design, Fabrication, Properties and Applications of Smart and Advanced Materials (illustrated ed.). CRC Press, 2016. p. 175. ISBN 1-4987-2249-0

- "Sodium" (PDF). Northwestern University. Archived from the original (PDF) on 2011-08-23. Retrieved 2011-11-21

- D'Andraba (1800). "Des caractères et des propriétés de plusieurs nouveaux minérauxde Suède et de Norwège, avec quelques observations chimiques faites sur ces substances". Journal de chimie et de physique. 51: 239

- Ivor L. Simmons (ed.). Applications of the Newer Techniques of Analysis. Springer Science & Business Media, 2012. p. 160. ISBN 1-4684-3318-0

- Various authors (1818). "The Quarterly journal of science and the arts" (PDF). The Quarterly Journal of Science and the Arts. Royal Institution of Great Britain. 5: 338. Retrieved 5 October 2010

- Lincoln, S. F.; Richens, D. T.; Sykes, A. G. (2004). "Metal Aqua Ions". Comprehensive Coordination Chemistry II. p. 515. doi:10.1016/B0-08-043748-6/01055-0. ISBN 978-0-08-043748-4

3

Various Compounds in Organometallic Chemistry

The p-Block metals can lose electrons easily and form organometallic compounds. Their reaction pattern includes oxidation, Lewis acidity, and its nucleophilic nature. The chapter closely examines the key concepts of organometallic chemistry to provide an extensive understanding of the subject.

P- Block Element

The reactions of organometallic compounds of electropositive elements are dominated by factors such as the carbanion character of the organic moiety and the availability of a coordination site on the central metal atom.

Reaction patterns:

a)   Oxidation

- All organometallic compounds are potentially reducing agents.

- Those of electropositive elements are in fact very strong reducing agents (many of them are pyrophoric in nature).

- The strong reducing character also presents a potential explosion hazard if the compounds are mixed with larger amount of oxidzing agents.

Why is it so?

All organometallic compounds of the electropositive metals that have unfilled valence orbitals, or that readily dissociate into fragments with unfilled orbitals are pyrophoric.

Examples:

$$Li_4(CH_3)_4,\ Zn(CH_3)_2,\ B(CH_3)_3\ and\ Al_2(CH_3)_6$$

Volatile pyrophoric compounds, such as $B(CH_3)_3$, may be handled in vacuum line and, inert atmosphere techniques are used for less volatile but air-sensitive compounds.

Compounds such as $Si(CH_3)_4$ and $Sn(CH_3)_4$ which do not have low-lying empty orbitals, require elevated temperatures to initiate combustion, and can be handled in air.

The combustion of many organometallic compounds takes place by a radical chain mechanism.

b)   Nucleophilic character

The partial negative charge of an organic group attached to an electropositive metal makes it a strong nucleophile and Lewis base.

This is referred to as its carbanion character even though the compound itself is not ionic.

Alkyllithium and alkylaluminium compounds and Grignard reagents are the most common carbanion reagents in laboratory-scale synthetic chemistry.

The carbanion character diminishes for the less metallic boron and silicon.

The carbanion character finds many synthetic applications.

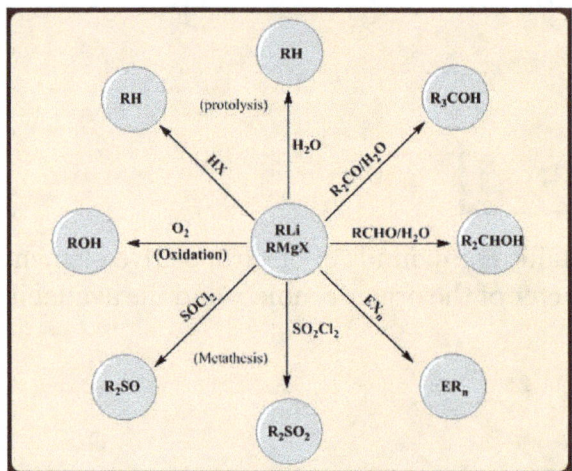

Where X = halide, E = B, Si, Ge, Sn, Pb, As and Sb

$$Al_2(CH_3)_6 + 6C_2H_5OH \rightarrow 2Al(OC_2H_5)_3 + 6CH_4$$

c)   Lewis acidity

Due to the presence of unoccupied orbitals on the metal atom, electron-deficient organometallic compounds are observed to be Lewis acids.

$$e.g. \ B(C_6H_5)_3 + LiC_6H_5 \rightarrow Li\left[B(C_6H_5)_4\right]$$

This reaction may be viewed as the transfer of the strong base C_6H_5-from the weak Lewis acid $Li^+$to the stronger acid B(III).

Organometallic species that are bridged by organic groups can also serve as Lewis acids and, in the process, bridge cleavage can take place.

$$Al_2(CH_3)_6 + 2N(C_2H_5)_3 \rightarrow 2(CH_3)_3 AlN(C_2H_5)_3$$

Electron deficient organometallic species are Lewis acids.

Organometallic Compounds of Boron and Aluminium

Organoboron Compounds

BMe3 is colorless, gaseous (b.p. -22 °C), and is monomeric. It is pyrophoric but not rapidly hydrolyzed by water.

Alkylboranes can be synthesized by metathesis between BX_3 and organometallic compounds of metals with low electronegativity, such as RMgX or AlR3.

$$BF_3 + 3CH_3MgBr \rightarrow B(CH_3)_3 + 3MgBrF \quad (solvent\ used;\ dibutyl\ ether)$$

Why dibutyl ether as a solvent: Has much lower vapor pressure than BMe_3 and as a result the separation by trap-to-trap distillation on a vacuum line is easy.

Also, there is a very weak association between BMe_3 and $OBu_2(Me_3B{:}OBu_2)$.

Although, trialkyl- and triarylboron compounds are mild Lewis acids, strong carbanion reagents lead to anions of the type $[BR_4]^-$.

Example, Na[BPh]$_4$: The bulky anion hydrolyses very slowly in neutral or basic water and is useful for the preparation of large positive cations.

$$Na[BPh]_4 + K^+ \rightarrow K[BPh]_4$$

K[BPh]$_4$ is insoluble, used for the gravimetric estimation (determination) of potassium, an example of the low solubility of large-cation and large-anion salts in water.

Organohaloboron compounds are more reactive than simple trialkylboron compounds.

Preparation:

$$2BCl_3 + 6Alr_3 \rightarrow 3R_2BCl + 6AlR_2Cl\ (metathesis)$$
$$2BCl_3 + BMe_3 \xrightarrow[diborane]{} 3BMeCl_2\ (redistribution\ reaction)$$

Reactions: (Protolysis reactions with ROH, R_2NH and other reagents)

$$3BMeCl_2 + 2HNR_2 \rightarrow BMe_2(NR2_2) + [R_2NH_2]Cl$$
$$BMe_2Cl + Li(C_4H_9) \rightarrow BMe_2(C_4H_9) + LiCl$$

Organoaluminium Compounds

With less bulky alkyl groups, dimerization occurs and one of the distinguishing features of alkyl bridge is the small Al-C-Al angle, which is ~ 75°.

The 3c,2e bonds are very weak and tend to dissociate in the pure liquid which increases with increase in the bulkiness of the alkyl group.

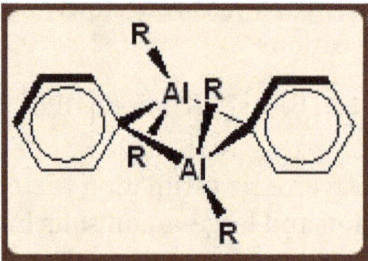

$$Al_2(CH_3)_6 \rightleftarrows 2Al(CH_3)_3 \quad K = 1.25 \times 10^{-8}$$

$$Al_2(C_2H_9)_6 \rightleftarrows 2Al(C_4H_9)_3 \quad K = 2.3 \times 10^{-4}$$

Perpendicular orientation of pheynl groups in Al2Ph6 Triphenylaluminium exists as a dimer with bridging η1-phenyl groups lying in a plane perpendicular to the line joining the two Al atoms.

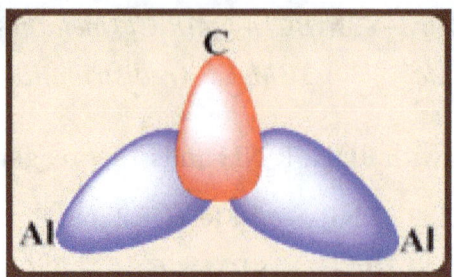

This structure is favored partly on steric grounds and partly by supplementation of the Al-C-Al bond by electron donation from the phenyl π-orbitals to the Al atoms.

Tendency for bridging: X > Ph > alkyl

3c,2e bonds formed by a symmetric combination of Al and C orbitals.

An additional interaction between the pπ orbital on C and an antisymmetric combination of Al orbitals.

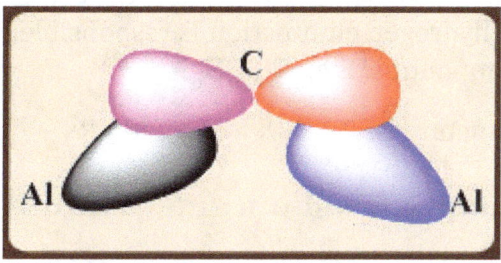

Synthesis

Very useful as alkene polymerization catalysts and chemical intermediates.

Expensive carbanion reagents for the replacement of halogens organic groups by metathesis.

Laboratory scale preparations involves:

$$2Al + 3Hg(CH_3)_2 \rightarrow Al_2(CH_3)_6 + 3Hg$$

Commercial method:

$$2Al + CH_3Cl \rightarrow Al_2Cl_2(CH_3)_4$$
$$Al_2Cl_2(CH_3)_4 + 6Na \rightarrow Al_2(CH_3)_6 + 2Al + 6NaCl$$

Commercial method for ethylaluminium and higher homologs:

$$2Al + 3H_2 + 6RHC = CH_2 \xrightarrow[110\ 200\ atm]{60\ 110} 2Al_2(CH_2CH_2R)_6$$

The reaction probably proceeds by the formation of a surface Al—H species that adds across the double bond of the alkene in a hydrometallation reaction.

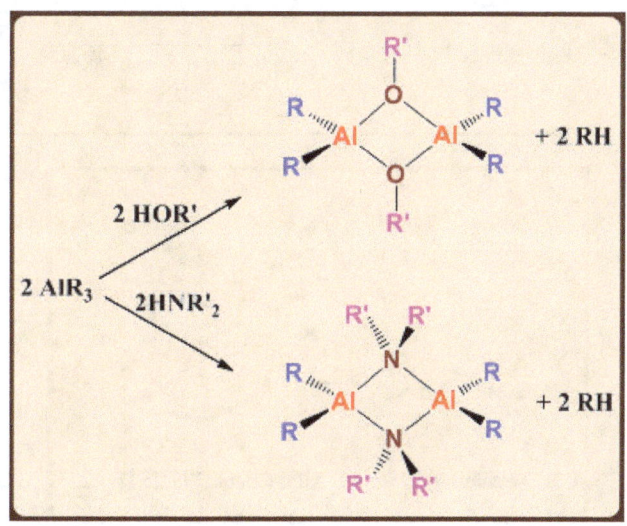

Reactions:

Alkylaluminum compounds are mild Lewis acids and form complexes with ethers, amines and anions. When heated, often β-hydrogen elimination is responsible for the decomposition of ethyl and higher alkylaluminium compounds. E.g. $Al(^iC_4H_9)_3$

Tendency towards bridging structure is: $PR_2^- > X^- > H^- > Ph^- > R^-$.

Organometallic Compounds of Gallium and Indium

$$3Li_4(C_2H_5)_4 + 4GaCl_3 \rightarrow 2LiCl + 4Ga(C_2H_5)_3$$

Trialkylgallium compounds are mild Lewis acids, so the corresponding metathesis reaction in ether produces the complex $(C_2H_5)_2OGa(C_2H_5)_3$. Similarly excess use of C_2H_5Li leads to the salt, $Li[Ga(C_2H_5)_4]$.

$$Li_4(C_2H_5)_4 + GaCl_3 \rightarrow 3LiCl + Li\big[Ga(C_2H_5)_4\big]$$

Alkylindium and alkylthalium compounds may be prepared similar to gallium analogs. $InMe_3$ is monomeric in the gas phase and in the solid the bond lengths indicate that association is very weak. Partial hydrolysis of $TlMe_3$ yields the linear $(MeTiMe]^+$ion, which is isoelectronic and iso-structural with $HgMe_2$.

CpIn and CoTl exist as monomers in the gas phase but are associated in solids {Inert-pair effect is displayed for In and Tl}. CpTl is useful as a synthetic reagent in organometallic chemistry because it is not as highly reducing as NaCp.

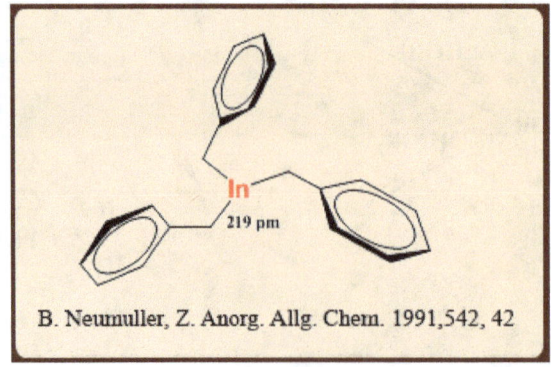

Species of the type R_4E_2(single E-E bond) and $[R_4E_2]^-$ (with E-E bond order of 1.5) can be prepared for Ga and In with bulky R groups (R = $(Me_3Si)_2CH$, 2,4,6-$^iPr_3C_6H_2$), and reduction of $[(2,4,6\text{-}^iPr_3C_6H_2)_4Ga_2]$ to $[(2,4,6\text{-}^iPr_3C_6H_2)_4Ga_2]^-$ is accompanied by a shortening of the Ga—Ga bond from 252-234 pm.

Using even bulkier substituents, it is possible to prepare gallium(I) compounds, RGa starting from GaI. No structural data are yet available for these monomers.

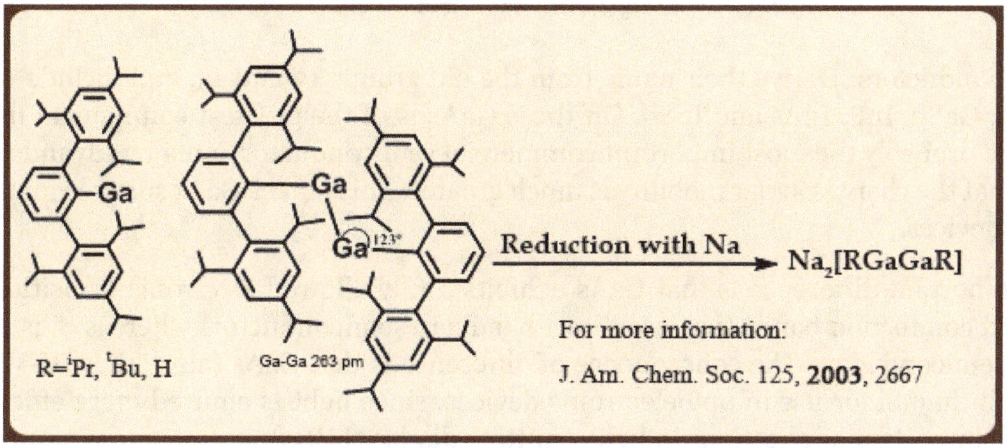

Ga
R=ⁱPr, ᵗBu, H
Ga
Ga
123°
Ga-Ga 263 pm
Reduction with Na
Na₂[RGaGaR]
For more information:
J. Am. Chem. Soc. 125, 2003, 2667

Crystallized as dimer but reverts to monomer when dissolved in cyclohexane.

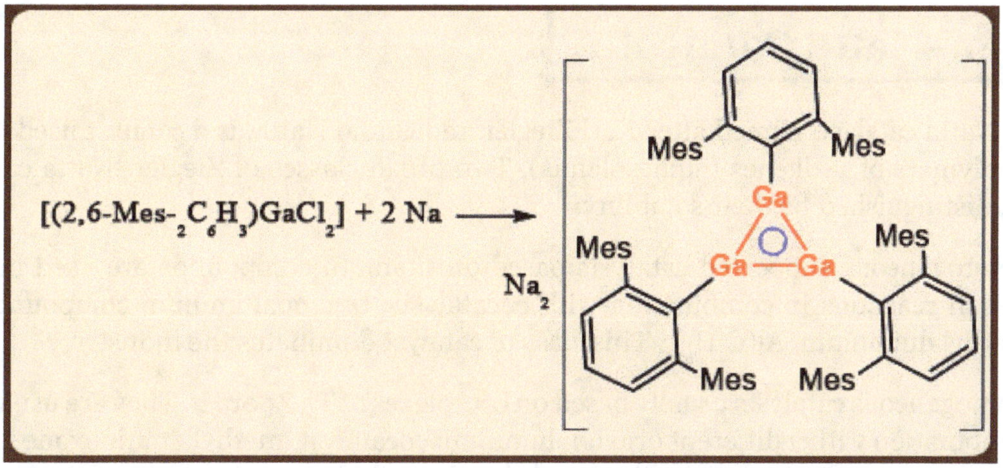

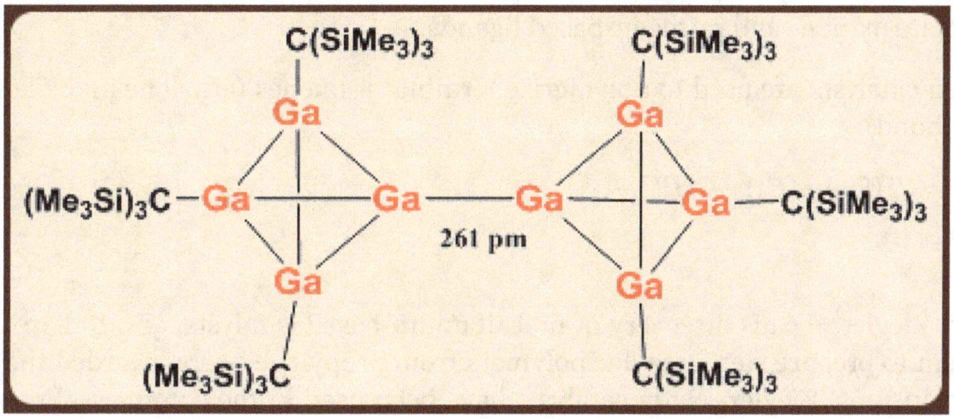

[(2,6-Mes-₂C₆H₃)GaCl₂] + 2 Na
Na₂
Mes
Mes
Ga
Mes
Mes
Ga
Ga
Mes
Mes

C(SiMe₃)₃
C(SiMe₃)₃
Ga
Ga
(Me₃Si)₃C
Ga
Ga
Ga
Ga
C(SiMe₃)₃
261 pm
Ga
Ga
(Me₃Si)₃C
C(SiMe₃)₃

Interest in organometallic comounds of Ga, In and Tl is mainly because of their potential use as precursors to semiconducting materials such as GaAs and InP. Volatile compounds can be used in the growth of thin films by MOCVD (metal organic chemical vapor deposition) or MOVPE (metal organic vapor phase epitaxy) techniques. Precursors include appropriate Lewis base adducts of metal alkyls, e.g. Me3Ga.NMe$_3$ and Me3In.PEt$_3$. Thermal decomposition of gaseous precursors result in semiconductors (III-V semiconductors) which can be deposited in thin films.

$$Me_3Ga(g) + AsH_3(g) \xrightarrow{1000-1150K} GaAs(s) + 3CH_4(g)$$

III-V semiconductors: Derive their name from the old groups 13 and 15, and include AlAs, AlSb, GaP, GaAs, GaSb, InP, InAs and InSb. Off these GaAs is of the greatest commercial interest. Although Si is probably the most important commercial semiconductor, a major advandage of GaAs over Si is that the charge carrier mobility is much greater. This makes GaAs suitable for high-speed electronic devices.

Another important difference is that GaAs exhibits a fuly allowed electronic transition between valence and conduction bands (i.e. it is direct band gap semiconductor) whereas Si is an indirect band gap semiconductor. The consequence of difference is that GaAs (also other III-V types) are more suited than Si for use in optoelectronic devices, since light is emitted more efficiently. The III-Vs have important applications in light-emitting diodes (LEDs).

Ziegler–Natta Catalyst

A Ziegler–Natta catalyst, named after Karl Ziegler and Giulio Natta, is a catalyst used in the synthesis of polymers of 1-alkenes (alpha-olefins). Two broad classes of Ziegler–Natta catalysts are employed, distinguished by their solubility:

- Heterogeneous supported catalysts based on titanium compounds are used in polymerization reactions in combination with cocatalysts, organoaluminum compounds such as triethylaluminium, $Al(C_2H_5)_3$. This class of catalyst dominates the industry.

- Homogeneous catalysts usually based on complexes of Ti, Zr or Hf. They are usually used in combination with a different organoaluminum cocatalyst, methylaluminoxane (or methylalumoxane, MAO). These catalysts traditionally include metallocenes but also feature multidentate oxygen- and nitrogen-based ligands.

Ziegler–Natta catalysts are used to polymerize terminal 1-alkenes (ethylene and alkenes with the vinyl double bond):

$$n\ CH_2=CHR \rightarrow -[CH_2-CHR]_n-;$$

History

German Karl Ziegler, for his discovery of first titanium-based catalysts, and Italian Giulio Natta, for using them to prepare stereoregular polymers from propylene, were awarded the Nobel Prize in Chemistry in 1963. Ziegler–Natta catalysts have been used in the commercial manufacture of various polyolefins since 1956. In 2010, the total volume of plastics, elastomers, and rubbers produced from alkenes with these and related (especially Phillips) catalysts worldwide exceeds 100 million tonnes. Together, these polymers represent the largest-volume commodity plastics as well as the largest-volume commodity chemicals in the world.

In the early 1950s workers at Phillips Petroleum discovered that chromium catalysts are highly effective for the low temperature polymerization of ethylene, which launched major industrial technologies. A few years later, Ziegler discovered that a combination of $TiCl_4$ and $Al(C_2H_5)_2Cl$ gave comparable activities for the production of polyethylene. Natta used crystalline α-$TiCl_3$ in combination with $Al(C_2H_5)_3$ to produce first isotactic polypropylene. Usually Ziegler catalysts refer to titanium-based systems for conversions of ethylene and Ziegler–Natta catalysts refer to systems for conversions of propylene. In the 1970s, magnesium chloride was discovered to greatly enhance the activity of the titanium-based catalysts. These catalysts were so active that the residual titanium was no longer removed from the product. They enabled to the commercialization of linear low-density polyethylene (LLDPE) resins and allowed the development of noncrystalline copolymers.

Also, in the 1960s, BASF developed a gas-phase, mechanically-stirred polymerization process for making polypropylene. In that process, the particle bed in the reactor was either not fluidized or not fully fluidized. In 1968, the first gas phase fluidized-bed polymerization process, the Unipol process, was commercialized by Union Carbide to produce polyethylene. In the mid-1980s, the Unipol process was further extended to produce polypropylene.

The features of the fluidized-bed process, including its simplicity and product quality, made it widely accepted all over the world. As of today, the fluidized-bed process is one of the two most widely used technologies for producing polypropylene.

Stereochemistry of Poly-1-alkenes

Natta first used polymerization catalysts based on titanium chlorides to polymerize propylene and other 1-alkenes. He discovered that these polymers are crystalline materials and ascribed their crystallinity to a special feature of the polymer structure called stereoregularity.

Short segments of polypropylene, showing examples of isotactic (above) and syndiotactic (below) tacticity.

The concept of stereoregularity in polymer chains is illustrated in the picture on the left with polypropylene. Stereoregular poly(1-alkene) can be isotactic or syndiotactic depending on the relative orientation of the alkyl groups in polymer chains consisting of units $-[CH_2-CHR]-$, like the CH_3 groups in the figure. In the isotactic polymers, all stereogenic centers CHR share the same configuration. The stereogenic centers in syndiotactic polymers alternate their relative configuration. A polymer that lacks any regular arrangement in the position of its alkyl substituents (R) is called atactic. Both isotactic and syndiotactic polypropylene are crystalline, whereas atactic polypropylene, which can also be prepared with special Ziegler–Natta catalysts, is amorphous. The stereoregularity of the polymer is determined by the catalyst used to prepare it.

Classes

Heterogeneous Catalysts

The first and dominant class of titanium-based catalysts (and some vanadium-based catalysts) for alkene polymerization can be roughly subdivided into two subclasses, (a) catalysts suitable for homopolymerization of ethylene and for ethylene/1-alkene copolymerization reactions leading to copolymers with a low 1-alkene content, 2–4 mol% (LLDPE resins), and (b) catalysts suitable for the synthesis of isotactic 1-alkenes. The overlap between these two subclasses is relatively small because the requirements to the respective catalysts differ widely.

Commercial catalysts are supported, i.e. bound to a solid with a high surface area. Both $TiCl_4$ and $TiCl_3$ give active catalysts. The support in the majority of the catalysts is $MgCl_2$. A third component of most catalysts is a carrier, a material that determines the size and the shape of catalyst particles. The preferred carrier is microporous spheres of amorphous silica with a diameter of 30–40 mm. During the catalyst synthesis, both the titanium compounds and $MgCl_2$ are packed into the silica pores. All these catalysts are activated with organoaluminum compounds such as $Al(C_2H_5)_3$.

All modern supported Ziegler–Natta catalysts designed for polymerization of propylene and higher 1-alkenes are prepared with $TiCl_4$ as the active ingredient and $MgCl_2$ as a support. Another component of all such catalysts is an organic modifier, usually an ester of an aromatic diacid or a diether. The modifiers react both with inorganic ingredients of the solid catalysts as well as with organoaluminum cocatalysts. These catalysts polymerize propylene and other 1-alkenes to highly crystalline isotactic polymers.

Homogeneous Catalysts

A second broad class of Ziegler–Natta catalysts are soluble in the reaction medium. Traditionally such homogeneous catalysts are derived from metallocenes but the structure of active catalysts have been significantly broadened.

Metallocene Catalysts

These catalysts are metallocenes together with a cocatalyst, typically MAO, $-[O-Al(CH_3)]_n-$. The idealized metallocene catalysts have the composition Cp_2MCl_2 (M = Ti, Zr, Hf) such as titanocene dichloride. Typically, the organic ligands are derivatives of cyclopentadienyl. In some complexes, the two cyclopentadiene (Cp) rings are linked with bridges, like $-CH_2-CH_2-$ or $>SiPh_2.$, Depending of the type of their cyclopentadienyl ligands, for example by using an *ansa*-bridge, metallocene catalysts can produce either isotactic or syndiotactic polymers of propylene and other 1-alkenes.

Non-metallocene Catalysts

Ziegler–Natta catalysts of the third class, non-metallocene catalysts, use a variety of complexes of various metals, ranging from scandium to lanthanoid and actinoid metals, and a large variety of ligands containing oxygen, nitrogen, phosphorus, and sulfur. The complexes are activated using MAO, as is done for metallocene catalysts.

Most Ziegler–Natta catalysts and all the alkylaluminium cocatalysts are unstable in air, and the alkylaluminium compounds are pyrophoric. The catalysts, therefore, are always prepared and handled under an inert atmosphere.

Mechanism of Ziegler–Natta Polymerization

The structure of active centers in Ziegler–Natta catalysts is firmly established only for metallocene catalysts. A metallocene complex Cp_2ZrCl_2 reacts with MAO and is transformed into a metallocenium ion $Cp_2Zr^+CH_3$. A polymer molecule grows in length by numerous insertion reactions of C=C bonds of 1-alkene molecules into the Zr–C bond in the ion:

$$Cp_2\overset{+}{Z}r-CH_3 + nCH_2 = CHR \ \rightarrow \ Cp_2\overset{+}{Z}r-(CH_2-CHR)_n - CH_3$$

Many thousands of alkene insertion reactions occur at each active center resulting in the formation of long polymer chains attached to the center. On occasion, the polymer chain is disengaged from the active centers in the chain termination reaction:

$$Cp_2\overset{+}{Z}r-(CH_2-CHR)_n-CH_3 + CH_2 = CHR \ \rightarrow \ Cp_2\overset{+}{Z}r-CH_2-CH_2R \ + \ CH_2 = CR-polymer$$

Another type of chain termination reaction called β-hydrogen elimination reaction also occurs periodically:

$$Cp_2\overset{+}{Z}r-(CH_2-CHR)_n-CH_3 \rightarrow \ Cp_2\overset{+}{Z}r-H \ + \ CH_2 = CR-polymer$$

Polymerization reactions of alkene with solid titanium-based catalysts occur at special titanium centers located on the exterior of the catalyst crystallites. Some titanium atoms in these crystallites react with organoaluminum cocatalysts with the formation of Ti–C bonds. The polymerization reaction of alkenes occurs similarly to the reactions in metallocene catalysts:

$$L_nTi-CH_2-CHR-polymer + CH_2=CHR \rightarrow L_nTi-CH_2-CHR-CH_2-CHR-polymer$$

The two chain termination reactions occurs quite rarely in Ziegler–Natta catalysis and the formed polymers have a too high molecular weight to be of commercial use. To reduce the molecular weight, hydrogen is added to the polymerization reaction:

$$L_nTi-CH_2-CHR-polymer + H_2 \rightarrow L_nTi-H + CH_3-CHR-polymer$$

The Cossee–Arlman mechanism describes the growth of stereospecific polymers. This mechanism states that the polymer grows through alkene coordination at a vacant site at the titanium atom, which is followed by insertion of the C=C bond into the Ti–C bond at the active center.

Zeigler Natta Polymerization Catalysts

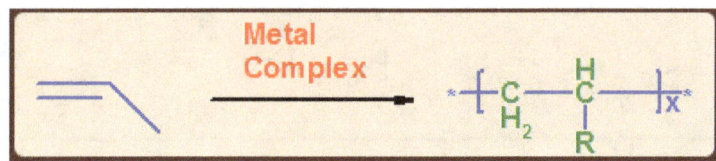

Insertion of aluminum alkyls into olefins was studied by Ziegler. During the systematic investigation of olefin polymerization, Ziegler realized that the most effective catalyst is the combination of $TiCl_4/AlEt_3$ which can polymerize ethylene at pressure as low as 1 bar. The application of Ziegler method to the polymerization of propylene and its establishment and the investigation of bulk properties was carried out by Natta and hence the methodology is called Ziegler-Natta process.

Important discovery: R_3Al + Lewis acids.

In the absence of reaction mechanism with solid proof, it is presumed that the reaction is due to the heterogeneous catalysis in which fibrous $TiCl_3$, alkylated on its surface is considered to be the active catalyst species.

Alkane Stereochemistry

Alkane stereochemistry concerns the stereochemistry of alkanes. Alkane conformers are one of the subjects of alkane stereochemistry.

Conformation of Alkane

Alkane conformers arise from rotation around sp³ hybridised carbon carbon sigma bonds. The smallest alkane with such a chemical bond, ethane, exists as an infinite number of conformations with respect to rotation around the C–C bond. Two of these are recognised as energy minimum (staggered conformation) and energy maximum (eclipsed conformation) forms. The existence of specific conformations is due to hindered rotation around sigma bonds, although a role for hyper-conjugation is proposed by a competing theory.

The importance of energy minimum and energy maximum is seen by extension of these concepts to more complex molecules for which stable conformations may be predicted as minimum energy forms. The determination of stable conformations has also played a large role in the establishment of the concept of asymmetric induction and the ability to predict the stereochemistry of reactions controlled by steric effects.

In the example of staggered ethane in Newman projection, a hydrogen atom on one carbon atom has a 60° torsional angle or torsion angle with respect to the nearest hydrogen atom on the other carbon so that steric hindrance is minimised. The staggered conformation is more stable by 12.5 kJ/mol than the eclipsed conformation, which is the energy maximum for ethane. In the eclipsed conformation the torsional angle is minimised.

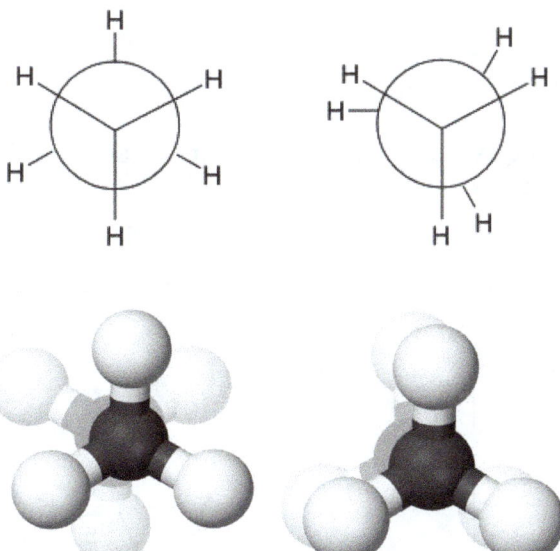

In butane, the two staggered conformations are no longer equivalent and represent two distinct conformers:the anti-conformation (left-most, below) and the gauche conformation (right-most, below).

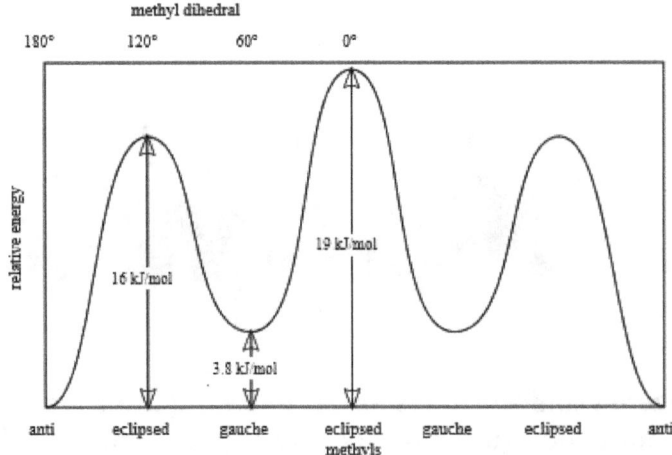

Both conformations are free of torsional strain, but, in the gauche conformation, the two methyl groups are in closer proximity than the sum of their van der Waals radii. The interaction between the two methyl groups is repulsive (van der Waals strain), and an energy barrier results.

A measure of the potential energy stored in butane conformers with greater steric hindrance than the 'anti'-conformer ground state is given by these values:

- Gauche, conformer – 3.8 kJ/mol

- Eclipsed H and CH_3 – 16 kJ/mol

- Eclipsed CH_3 and CH_3 – 19 kJ/mol.

The eclipsed methyl groups exert a greater steric strain because of their greater electron density compared to lone hydrogen atoms.

Relative energies of conformations of butane with respect to rotation of the central C-C bond.

The textbook explanation for the existence of the energy maximum for an eclipsed conformation in ethane is steric hindrance, but, with a C-C bond length of 154 pm and a Van der Waals radius for hydrogen of 120 pm, the hydrogen atoms in ethane are never in each other's way. The question of whether steric hindrance is responsible for the eclipsed energy maximum is a topic of debate to this day. One alternative to the steric hindrance explanation is based on hyperconjugation as analyzed within the Natural Bond Orbital framework. In the staggered conformation, one C-H sigmabonding orbital donates electron density to the antibonding orbital of the other C-H bond. The energetic stabilization of this effect is maximized when the two orbitals have maximal overlap, occurring in the staggered conformation. There is no overlap in the eclipsed conformation, leading to a disfavored energy maximum. On the other hand, an analysis within quantitative molecular orbital theory shows that 2-orbital-4-electron (steric) repulsions are dominant over hyperconjugation. A valence bond theory study also emphasizes the importance of steric effects.

Nomenclature

Naming alkanes per standards listed in the IUPAC Gold Book is done according to the Klyne–Prelog system for specifying angles (called either torsional or dihedral angles) between substituents around a single bond:

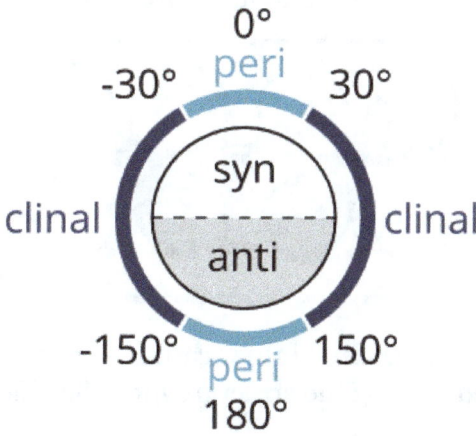

- a torsion angle between 0° and ± 90° is called syn (s)

- a torsion angle between ± 90° and 180° is called anti (a)

- a torsion angle between 30° and 150° or between −30° and −150° is called clinal

- a torsion angle between 0° and ± 30° or ± 150° and 180° is called periplanar (p)

- a torsion angle between 0° and ± 30° is called synperiplanar or syn- or cis-conformation (sp)

- a torsion angle between 30° to 90° and −30° to −90° is called synclinal or gauche or skew (sc)

- a torsion angle between 90° and 150° or −90° and −150° is called anticlinal (ac)

- a torsion angle between ± 150° and 180° is called antiperiplanar or anti- or trans-conformation (ap).

Torsional strain results from resistance to twisting about a bond.

Special Cases

In *n*-pentane, the terminal methyl groups experience additional pentane interference.

Replacing hydrogen by fluorine in polytetrafluoroethylene changes the stereochemistry from the zigzag geometry to that of a helix due to electrostatic repulsion of the fluorine atoms in the 1,3 positions. Evidence for the helix structure in the crystalline state is derived from X-ray crystallography and from NMR spectroscopy and circular dichroism in solution.

Organosilicon and Organogermanium Compounds

Organosilicon compounds are extensively studied due to the wide range of commercial applications as water repellents, lubricants, and sealants.

Many oxo-bridged organosilicon compounds can be synthesized.

e.g. $(CH_3)_3Si—O—Si(CH_3)_3$ which is resistant to moisture and air.

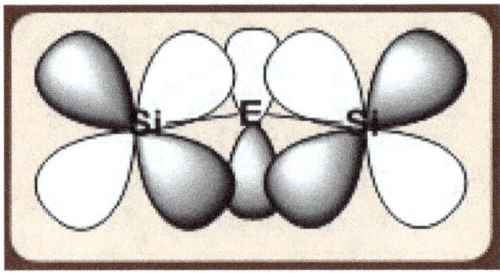

The lone pairs on O are partially delocalized into vacant σ^*- orbitals of Si, as a result the directionality of the Si-O bond is reduced making the structure more flexible.

This flexibility permits silicone elastomers to remain rubber-like down to very low temperature.

Delocalization also accounts for low basicity of an O atom attached to silicon as the electrons needed for the O atom to act as a base are partially removed.

The planarity of $N(SiH_3)_3$ is also explained by the delocalization of the lone pair on N which makes it very weakly basic.

$$nMeCl + Si / Cu \rightarrow Me_nSiCl_{4-n} \quad (temp\ 573\ K)$$
$$SiCl_4 + 4RLi \rightarrow R_4Si$$
$$SiCl_4 + RLi \rightarrow RSiCl_3$$
$$SiCl_4 + 2RMgCl \rightarrow R_2SiCl_2 + 2MgCl_2$$
$$Me_2SiCl_2 + {}^tBuLi \rightarrow {}^t BuMe_2SiCl + LiCl$$

Si—C bonds are relatively strong (bond enthalpy is 318 kJ mol^{-1}) and R_4Si derivatives possess high thermal stabilities.

Et4Si on chlorination gives $(ClCH_2CH_2)_4Si$, in contrast to the chlorination of R_4Ge or R_4Sn which yields R_nGeCl_{4-n} or $R_nSn_nCl_{4-n}$.

Me_2SiCl_2 on hydrolysis produce silicones.

$$Me_3SiCl + NaCp \rightarrow (n^1 - Cp)SiHMe_3$$

$(\eta^1\text{-}C_5Me_5)_2SiBr_2$ on treatment with anthracene/potassium gives $Cp*_2Si$

Solid state structure of $Cp*_2Si$ consists of two independent molecules which differ in the relative orientations of the Cp rings.

In one molecule, they are parallel and staggered whereas in the other, they are tilted with an angle of 167°at Si.

The reaction between R_2SiCl_2 and alkali metal or alkali naphthalides give cyclo-$(R_2Si)_n$ by loss of Cl- and Si—Si bond formation.

Bulky R groups favour small rings [e.g. $(2,6\text{-}Me_2C_6H_3)_6Si_3$ and $^tBu_6Si_3$] while smaller R groups encourage the formation of large rings [$Me_{12}Si_6$, $Me_{14}Si_7$ and $Me_{32}Si_{16}$]

$$Ph_2SiCl_2 + Li(SiPh_2)_5Li \rightarrow cyclo - Ph_{12}Si_6 + 2LiCl$$

Bulky substituents stabilize $R_2Si=SiR_2$ compounds. The sterically demanding $2,4,6\text{-}^iPr_3C_6H_2$ provided first example of compound containing conjugated Si=Si bonds.

Has s-cis configuration in both solution and the solid state.

Similar Germanium Compounds are also Known

*The spatial arrangement of two conjugated double bonds about the intervening single bond is described as s- cis if synperiplanar and s-trans if antiperiplanar.

Organotin Chemistry

Organotin compounds or stannanes are chemical compounds based on tin with hydrocarbon substituents. Organotin chemistry is part of the wider field of organometallic chemistry. The first organotin compound was diethyltin diiodide ($(C_2H_5)_2SnI_2$), discovered by Edward Frankland in 1849. The area grew rapidly in the 1900s, especially after the discovery of the Grignard reagents, which are useful for producing Sn-C bonds. The area remains rich with many applications in industry and continuing activity in the research laboratory.

Organotin compounds are those with tin linked to hydrocarbons.

Structure of Organotin Compounds

Organotin compounds are generally classified according to their oxidation states. Tin(IV) compounds are much more common and more useful.

Organic Derivatives of Tin(IV)

The tetraorgano derivatives are invariably tetrahedral. Compounds of the type SnRR'RR' have been resolved into individual enantiomers.

Organotin Halides

Organotin chlorides have the formula $R_{4-n}SnCl_n$ for values of n up to 4. Bromides, iodides, and fluorides are also known but less important. These compound are known for many R groups. They are always tetrahedral. The tri- and dihalides form adducts with good Lewis bases such as pyridine. The fluorides tend to associate such that dimethyltin difluoride forms sheet-like polymers. Di- and especially triorganotin halides, e.g. tributyltin chloride, exhibit toxicities approaching that of hydrogen cyanide.

Organotin Hydrides

Organotin hydrides have the formula $R_{4-n}SnH_n$ for values of n up to 4. The parent member of this series, stannane (SnH_4), is an unstable colourless gas. Stability is correlates with the number of organic substituents. Tributyltin hydride is used as a source of hydride radical in some organic reactions.

Organotin Oxides and Hydroxides

Organotin oxides and hydroxides are common products from the hydrolysis of organotin halides. Unlike the corresponding derivatives of silicon and germanium, tin oxides and hydroxides often adopt structures with penta- and even hexacoordinated tin centres, especially for the diorgano- and monoorgano derivatives. The group Sn-O-Sn is called a stannoxane. Structurally simplest of the oxides and hydroxides are the triorganotin derivatives. A commercially important triorganotin hydroxides is the acaricide Cyhexatin (also called Plictran), $(C_6H_{11})_3SnOH$. Such triorganotin hydroxides exist in equilibrium with the distannoxanes:

$$2\,R_3SnOH \rightleftharpoons R_3SnOSnR_3 + H_2O$$

With only two organic substituents on each Sn centre, the diorganotin oxides and hydroxides are structurally more complex than the triorgano derivatives. The simple geminal diols $(R_2Sn(OH)_2)$ and monomeric stannanones $(R_2Sn{=}O)$ are unknown. Diorganotin oxides (R_2SnO) are polymers except when the organic substituents are very bulky, in which case cyclic trimers or, in the case of $R = CH(SiMe_3)_2$ dimers, with Sn_3O_3 and Sn_2O_2 rings. The distannoxanes exist as dimers of dimers with the formula $[R_2SnX]_2O_2$ wherein the X groups (e.g., chloride, hydroxide, carboxylate) can be terminal or bridging. The hydrolysis of the monoorganotin trihalides has the potential to generate stannanoic acids, $RSnO_2H$. As for the diorganotin oxides/hydroxides, the monoorganotin species form structurally complex because of the occurrence of dehydration/hydration, aggregation. Illustrative is the hydrolysis of butyltin trichloride to give $[(BuSn)_{12}O_{14}(OH)_6]^{2+}$.

Idealized structure of trimeric diorganotin oxide.

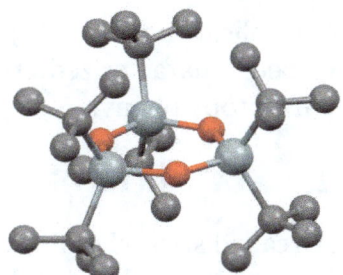

Ball-and-stick model for $(t\text{-}Bu_2SnO)_3$.

Structure of diorganotin oxide, highlighting the extensive intermolecular bonding.

Hypercoordinated Stannanes

Unlike carbon(IV) analogues but somewhat like silicon compounds, tin(IV) can also be coordinated to five and even six atoms instead of the regular four. These hypercoordinated compounds usually have electronegative substituents. Numerous examples of hypervalency are provided by the organotin oxides and associated carboxylates and related pseudohalide derivatives. The organotin halides for adducts, e.g. Me_2SnCl_2(bipyridine.

The all-organic penta- and hexaorganostannates have even been characterized, while in the subsequent year a six-coordinated tetraorganotin compound was reported. A crystal structure of room-temperature stable (in argon) all-carbon pentaorganostannane was reported as the lithium salt with this structure:

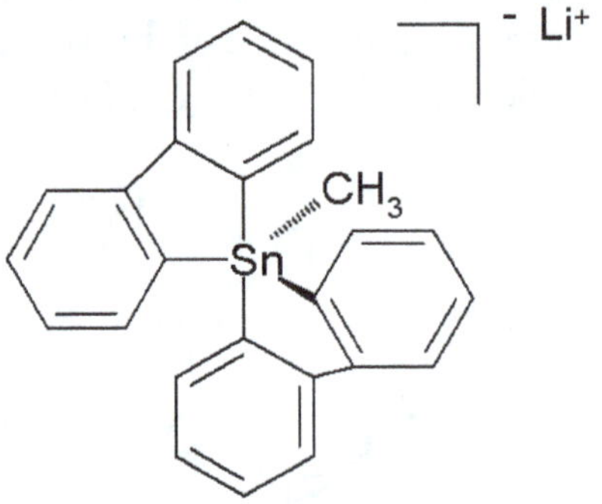

In this distorted trigonal bipyramidal structure the carbon to tin bond lengths (2.26 Å apical, 2.17 Å equatorial) are larger than regular C-Sn bonds (2.14 Å) reflecting its hypervalent nature.

Triorganotin Cations

Some reactions of triorganotin halides implicate a role for R3Sn+ intermediates. Such cations are analogous to carbocations. They have been characterized crystallographically when the organic substituents are large, such as 2,4,6-triisopropylphenyl.

Tin Radicals (Organic Derivatives of Tin(III))

Tin radicals, with the formula R_3Sn, are called stannyl radicals. They are invoked as intermediates in certain atom-transfer reactions. For example, tributyltin hydride (tri-n-butylstannane) serves as a useful source of "hydrogen atoms" because of the stability of the tributytin radical.

Organic Derivatives of Tin(II)

Organotin(II) compounds are somewhat rare. Compounds with the empirical formula SnR_2 are somewhat fragile and exist as rings or polymers when R is not bulky. The polymers, called polystannanes, have the formula $(SnR_2)_n$.

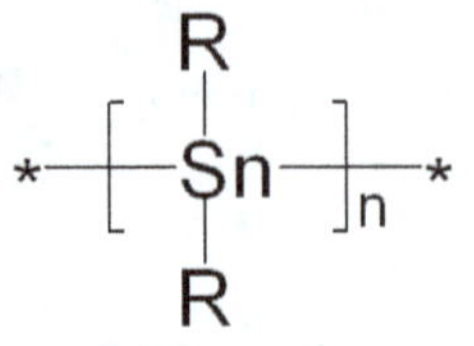

In principle divalent tin compounds might be expected to form analogues of alkenes with a formal double bond. Indeed, compounds with the formula Sn_2R_4, called distannenes, are known for certain organic substituents. The Sn centres tend to be highly pyramidal. Monomeric compounds with the formula SnR_2, analogues of carbenes are also known in a few cases. One example is Sn(-$SiR_3)_2$, where R is the very bulky $CH(SiMe_3)_2$ (Me = methyl). Such species reversibly dimerize to the distannylene upon crystallization:

$$2\ R_2Sn \rightleftharpoons (R_2Sn)_2$$

Stannenes, compounds with tin–carbon double bonds, are exemplified by derivatives of stannabenzene. Stannoles, structural analogs of cyclopentadiene, exhibit little C-Sn double bond character.

Organic Derivatives of Tin(I)

Compounds of Sn(I) are rare and only observed with very bulky ligands. One prominent family of cages is accessed by pyrolysis of the 2,6-diethylphenyl-substituted tristannylene [Sn(C_6H_3-2,6-$Et_2)_2]_3$, which affords the cubane-type cluster and a prismane. These cages contain Sn(I) and have the formula [Sn(C_6H_3-2,6-Et_2)]$_n$ where n = 8, 10. A stannyne contains a carbon to tin triple bond and a distannyne a triple bond between two tin atoms (RSnSnR). Distannynes only exist for extremely bulky substituents. Unlike alkynes, the C-Sn-Sn-C core of these distannynes are nonlinear, although they are planar. The Sn-Sn distance is 3.066(1) Å, and the Sn-Sn-C angles are 99.25(14)°. Such compounds are prepared by reduction of bulky aryltin(II) halides.

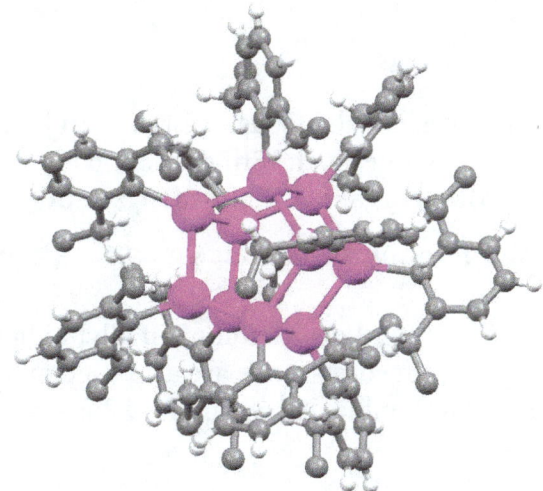

Structure of an $Ar_{10}Sn_{10}$ "prismane", a compound containing Sn(I) (Ar = 2,6-diethylphenyl).

Preparation of Organotin Compounds

Organotin compounds can be synthesised by numerous methods. Classic is the reaction of a Grignard reagent with tin halides for example tin tetrachloride. An example is provided by the synthesis of tetraethyltin:

$$4\ EtMgBr + SnCl_4 \rightarrow Et_4Sn + 4\ MgClBr$$

The symmetrical tetraorganotin compounds, especially tetraalkyl derivatives, can then be converted to various mixed chlorides by redistribution reactions (also known as the "Kocheshkov comproportionation" in the case of organotin compounds):

$$3\ R_4Sn + SnCl_4 \rightarrow 4\ R_3SnCl$$

$$R_4Sn + SnCl_4 \rightarrow 2\ R_2SnCl_2$$

$$R_4Sn + 3\ SnCl_4 \rightarrow 4\ RSnCl_3$$

A related method involves redistribution of tin halides with organoaluminium compounds.

The mixed organo-halo tin compounds can be converted to the mixed organic derivatives, as illustrated by the synthesis of dibutyldivinyltin:

$$Bu_2SnCl_2 + 2\ C_2H_3MgBr \rightarrow Bu_2Sn(C_2H_3)_2 + 2\ MgBrCl$$

The organotin hydrides are generated by reduction of the mixed alkyl chlorides. For example, treatment of dibutyltin dichloride with lithium aluminium hydride gives the dibutyltin dihydride, a colourless distillable oil:

$$Bu_2SnCl_2 + \frac{1}{2}LiAlH_4 \rightarrow Bu_2SnH_2 + \frac{1}{2}LiAlCl_4$$

The Wurtz-like coupling of alkyl sodium compounds with tin halides yields tetraorganotin compounds.

Reactions of Organotin Compounds

Important reactions, discussed above, usually focus on organotin halides and pseudohalides with nucleophiles. In the area of organic synthesis, the Stille reaction is considered important. It entails coupling reaction with sp2-hybridized organic halides catalyzed by palladium:

$$R-X+R'-SnR''3 \xrightarrow{Pd\ catalyst} R-R'+XSnR''3$$

and organostannane additions (nucleophilic addition of an allyl-, allenyl-, or propargylstannanes to an aldehydes and imines). Organotin compounds are also used extensively in radical chemistry (e.g. radical cyclizations, Barton–McCombie deoxygenation, Barton decarboxylation, etc.).

Applications

An organotin compound is commercially applied as stabilizers in polyvinyl chloride. In this capacity, they suppress degradation by removing allylic chloride groups and by absorbing hydrogen chloride. This application consumes about 20,000 tons of tin each year. The main class of organotin compounds are diorganotin dithiolates with the formula $R_2Sn(SR')_2$. The Sn-S bond is the reactive component. Diorganotin carboxylates, e.g., dibutyltin dilaurate, are used as catalysts for the formation of polyurethanes, for vulcanization of silicones, and transesterification.

n-Butyltin trichloride is used in the production of tin dioxide layers on glass bottles by chemical vapor deposition.

Biological Applications

"Tributyltins" are used as industrial biocides, e.g. as antifungal agents in textiles and paper, wood pulp and paper mill systems, breweries, and industrial cooling systems. Triphenyltin derivativess are used as active components of antifungal paints and agricultural fungicides. Other triorgano-tins are used as miticides and acaricides. Tributyltin oxide has been extensively used as a wood preservative.

Tributyltin compounds were once widely used as marine anti-biofouling agents to improve the efficiency of ocean-going ships. Concerns over toxicity of these compounds (some reports describe biological effects to marine life at a concentration of 1 nanogram per liter) led to a worldwide ban by the International Maritime Organization.

Organotin complexes have been studied in anticancer therapy.

- Organotin compounds

Tetrabutyltin colorless oil, precursor
to the other butyl-tin compounds

Tributyltin oxide, a colorless to pale yellow
liquid used in wood preservation

Triphenyltin acetate, an off-white crystalline solid,
used as an insecticide and a fungicide

Triphenyltin chloride, a highly toxic white solid,
used as a biocide

Trimethyltin chloride, a toxic white solid, once used as a biocide

Triphenyltin hydroxide, an off-white powder, used as a fungicide

Azocyclotin, a white solid, used as a long-acting acaricide
for control of spider mites on plants

Cyhexatin, a white solid, used as an acaricide and miticide

Hexamethylditin used as an intermediate in chemical synthesis

Tetraethyltin, boiling point 63–65° /12 mm is a catalyst

Toxicity

Triorganotin compounds can be highly toxic. Tri-*n*-alkyltins are phytotoxic and therefore cannot be used in agriculture. Depending on the organic groups, they can be powerful bactericides and fungicides. Reflecting their high bioactivity, "tributyltins" were once used in marine anti-fouling paint.

Tetraorgano-, diorgano-, and monoorganotin compounds generally exhibit low toxicity and low biological activity. DBT may however be immunotoxic.

Organolead Compound

Organolead compounds are chemical compounds containing a chemical bond between carbon and lead. Organolead chemistry is the corresponding science. The first organolead compound was hexaethyldilead ($Pb_2(C_2H_5)_6$), first synthesized in 1858. Sharing the same group with carbon, lead is tetravalent.

Going down the carbon group the C–X (X = C, Si, Ge, Sn, Pb) bond becomes weaker and the bond length larger. The C–Pb bond in tetramethyllead is 222 pm long with a dissociation energy of 49 kcal/mol (204 kJ/mol). For comparison the C–Sn bond in tetramethyltin is 214 pm long with dissociation energy 71 kcal/mol (297 kJ/mol). The dominance of Pb(IV) in organolead chemistry is remarkable because inorganic lead compounds tend to have Pb(II) centers. The reason is that with inorganic lead compounds elements such as nitrogen, oxygen and the halides have a much higher electronegativity than lead itself and the partial positive charge on lead then leads to a stronger contraction of the 6s orbital than the 6p orbital making the 6s orbital inert; this is called the inert pair effect.

By far the most important organolead compound is tetraethyllead, formerly used as an anti-knocking agent. The most important lead reagents for introducing lead are lead tetraacetate and lead chloride.

The use of organoleads is limited partly due to their toxicity, although the toxicity is only 10% of that of palladium compounds.

Synthesis

Organolead compounds can be derived from Grignard reagents and lead chloride. For example, methylmagnesium chloride reacts with lead chloride to tetramethyllead, a water-clear liquid with boiling point 110 °C and density 1.995 g/cm³. Reaction of a lead(II) source with sodium cyclopentadienide gives the lead metallocene, plumbocene.

Certain arene compounds react directly with lead tetraacetate to aryl lead compounds in an electrophilic aromatic substitution. For instance anisole with lead tetraacetate forms 'p-methoxyphenyllead triacetate in chloroform and dichloroacetic acid:

Other compounds of lead are organolead halides of the type $R_nPbX_{(4-n)}$, organolead sulfinates ($R_{n-}Pb(OSOR)_{(4-n)}$) and organolead hydroxides ($R_nPb(OH)_{(4-n)}$). Typical reactions are:

$$R_4Pb + HCl \rightarrow R_3PbCl + RH$$

$$R_4Pb + SO_2 \rightarrow R_3PbO(SO)R$$

$$R_3PbCl + 1/2 Ag_2O\ (aq) \rightarrow R_3PbOH + AgCl$$

$$R_2PbCl_2 + 2\ OH^- \rightarrow R2Pb(OH)2 + 2\ Cl^-$$

$R_2Pb(OH)_2$ compounds are amphoteric. At pH lower than 8 they form R_2Pb^{2+} ions and with pH higher than 10, $R_2Pb(OH)_3^-$ ions.

Derived from the hydroxides are the plumboxanes:

$$2\ R_3PbOH + Na \rightarrow (R_3Pb)_2O + NaOH + 1/2\ H_2$$

which give access to polymeric alkoxides:

$$(R_3Pb)_2O + R'OH \rightarrow 1/n\ (R_3PbOR')_n - n\ H_2O$$

Reactions

The C–Pb bond is weak and for this reason homolytic cleavage of organolead compounds to free radicals is easy. In its anti-knocking capacity, its purpose is that of a radical initiator. General reaction types of aryl and vinyl organoleads are transmetalation for instance with boronic acids and acid-catalyzed heterocyclic cleavage. Organoleads find use in coupling reactions between arene compounds. They are more reactive than the likewise organotins and can therefore be used to synthesise sterically crowdedbiaryls.

In oxyplumbation, organolead alkoxides are added to polar alkenes:

$$H_2C=CH-CN + (Et_3PbOMe)_n \rightarrow MeO-CH_2-HC(PbEt_3)-CN \rightarrow MeO-CH_2-CH_2-CN$$

The alkoxide is regenerated in the subsequent methanolysis and, therefore, acts as a catalyst.

Aryllead Triacetates

The lead substituent in p-methoxyphenyllead triacetate is displaced by carbon nucleophiles such as the phenol2,4,6-*trimethylphenol* (mesitol) exclusively at the aromatic ortho position:

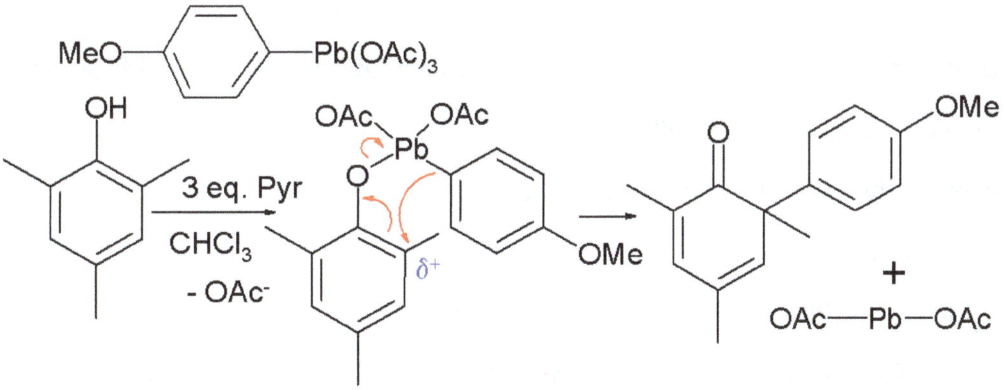

The reaction requires the presence of a large excess of a coordinating amine such as pyridine which presumably binds to lead in the course of the reaction. The reaction is insensitive to radical scavengers and therefore a free radical mechanism can be ruled out. The reaction mechanism is likely to involve nucleophilic displacement of an acetate group by the phenolic group to a diorganolead intermediate which in some related reactions can be isolated. The second step is then akin to a Claisen rearrangement except that the reaction depends on the electrophilicity (hence the ortho preference) of the phenol.

The nucleophile can also be the carbanion of a β-dicarbonyl compound:

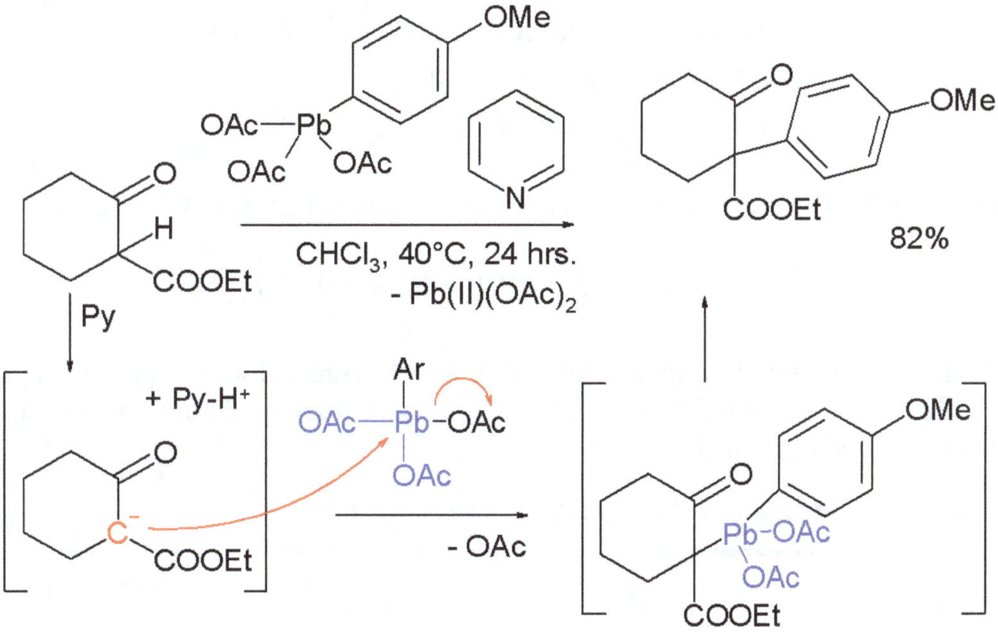

The carbanion forms by proton abstraction of the acidic α-proton by pyridine (now serving a double role) akin to the Knoevenagel condensation. This intermediate displaces an acetate ligand to a diorganolead compound and again these intermediates can be isolated with suitable reactants as unstable intermediates. The second step is reductive elimination with formation of a new C–C bond and lead(II) acetate.

Reactive Intermediates

Organolead compounds form a variety of reactive intermediates such as lead free radicals:

$$Me_3PbCl + Na\ (77\ K) \rightarrow Me_3Pb\cdot$$

and plumbylenes, the lead carbene counterparts:

$$Me_3Pb\text{-}Pb\text{-}Me_3 \rightarrow [Me_2Pb]$$

$$[Me_2Pb] + (Me_3Pb)_2 \rightarrow Me_3Pb\text{-}Pb(Me)_2\text{-}PbMe_3$$

$$Me_3Pb\text{-}Pb(Me)_2\text{-}PbMe_3 \rightarrow Pb(0) + 2\ Me_4Pb$$

These intermediates break up by disproportionation.

Plumbylidines of the type RPb (formally Pb(I)) are ligands to other metals in L_nMPbR compounds (compare to carbon metal carbynes).

Organotin and Organolead Compounds

Preparation of Sn(IV) Derivatives

$$3SnCl_4 + 4R_3Al \xrightarrow{R_2'O} 3R_4Sn + 4AlCl_3$$

$$R_4Sn + SnCl_4 \rightarrow R_3SnCl + RSnCl_3 \xrightarrow{500K} 2R_2SnCl_2$$

$$SnCl_4 + 4RMgBr \rightarrow R_4Sn + 4MgBrCl$$

$$SnCl_2 + Ph_2Hg \rightarrow Ph_2SnCl_2 + Hg$$

Tin(II) organometallics of the type R_2Sn, containing Sn-C σ-bonds, are stabilized only if R is sterically demanding.

$$SnCl_2 + 2Li\big|(Me_3Si)_2\,CH\big| \rightarrow \big\{(Me_3Si)_2\,CH\big\}_2\,Sn$$

(monomeric in solution and dimeric in solid state). But the dimer does not possess a planar Sn_2R_4 framework unlike an analogous alkene, and Sn—Sn bond distance (267 pm) is shorter than a normal Sn—Sn single bond (276 pm).

Sn_2R_4 has a trans bent structure with a weak Sn=Sn double bond

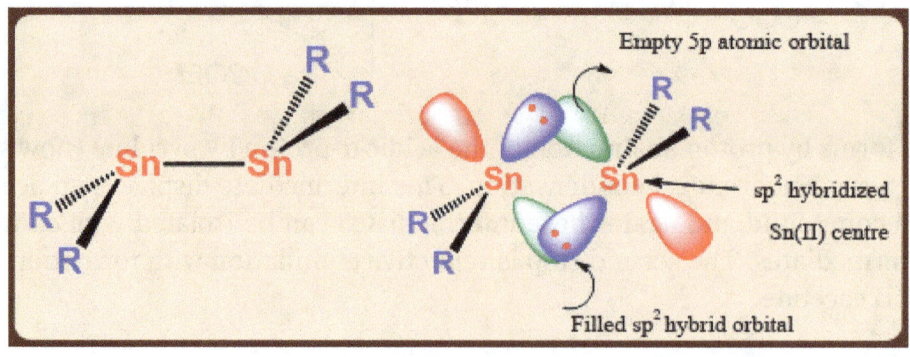

Look into the reactions of R_3SnCl with various reagents to form useful tin containing starting materials
The first organotin(II) hydride was reported only in 2000.

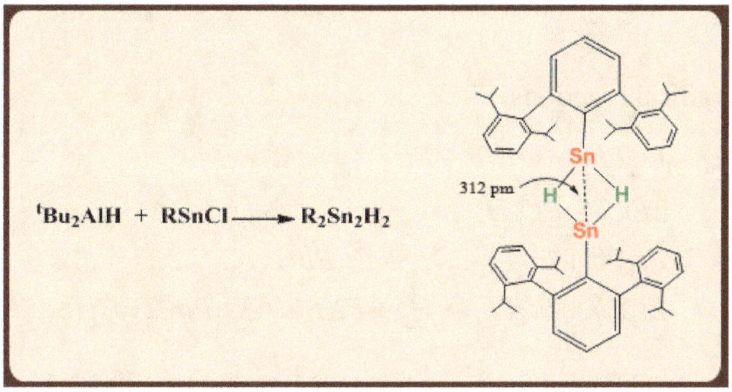

$$^tBu_2AlH + RSnCl \longrightarrow R_2Sn_2H_2$$

Shows dimeric structure in the solid state containing hydride bridges (Sn-Sn = 312 pm).

Commercial uses and Environmental Problems

Organotin(II) compounds find wide range of applications due to their catalytic and biocidal properties.

$^nBu_3SnOAc$ is an effective fungicide and bactericide and also a polymerization catalyst.

$^nBu_2Sn(OAc)_2$ is used as a polymerization catalyst and a stabilizer for PVC.

$^nBu_3SnOSn^nBu_3$ is algicide, fungicide and wood-preserving agent.

$^nBu_3SnCl$ is a bactericide and fungicide.

Ph_3SnOH used as an agricultural fungicide for crops such as potato, sugar beet and peanuts.

The cylic compound $(^nBu_2SnS)_3$ is used as a stabilizer for PVC.

Tributyltin derivatives have been used as antifouling agents, applied to the underside of ships' hulls to prevent the build-up of, for example, barnacles.

Global legislation now bans or greatly restricts the use of organotin-based anti-fouling agents on environmental grounds. Environmental risks associated with the uses of organotin compounds as pesticides, fungicides and PVC stabilizers are also a cause for concern.

*A barnacle is a type of arthropod belonging to infraclass Cirripedia in the sub-phylum Crustacea, and is hence related to crabs and lobsters.

Organolead Compounds

Tetraethyllead

$$4NaPb + 4EtCl \rightarrow Et_4Pb + 3Pb + 4NaCl \ [at\ 373\ K\ in\ an\ autoclave]$$

Laboratory Scale,

$$2PbCl_2 + 4RMgBr \xrightarrow[-4MgBrCl]{Et_2O} 2\{R_2Pb\} \rightarrow R_4Pb + Pb$$

Thermolysis leads to radical reactions.

$$Et_4Pb \rightarrow Et_3Pb. + Et.$$

$$2Et. \rightarrow n^-C_4H_{10}$$

$$Et_3Pb. + Et. \rightarrow C_2H_4 + Et_3PbH$$

$$Et_3Pb. + Et_4Pb \rightarrow H_2 + Et_3Pb. + Et_3PbCH_2CH_2$$

Tetraalkyl and tetraaryl lead compounds are inert with respect to attack by air and water at room temperature.

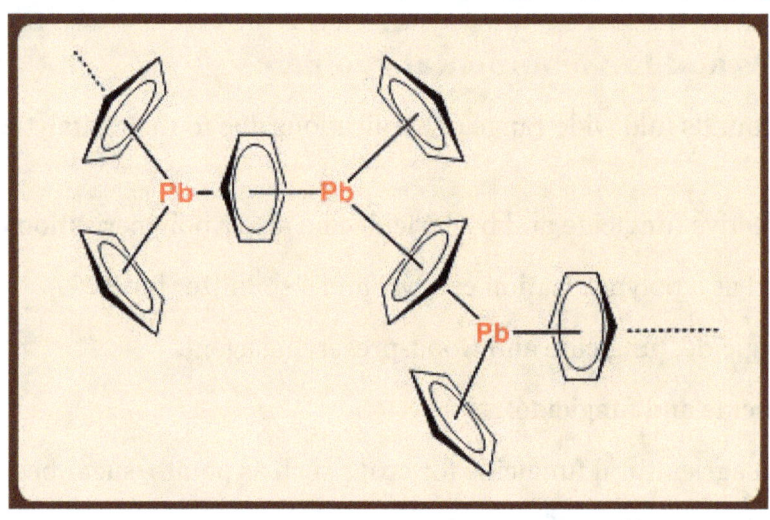

Me$_3$PbCl consists of linear chain

Solid state structure of Cp$_2$Pb shows polymeric nature, but in the gas phase, discrete Cp$_2$Pb molecules are present which possess the bent structure similar to silicon analogue.

R$_2$Pb=PbR$_2$ are similar to analogues tin compounds.

References

- Aue, W. P. and Bartholdi, E. and Ernst, R. R., Two-dimensional spectroscopy. Application to nuclear magnetic resonance; The Journal of Chemical Physics, 64, 2229-2246 (1976)

- Hoff, Ray; Mathers, Robert T., eds. (2010). Handbook of Transition Metal Polymerization Catalysts (Online ed.). John Wiley & Sons. doi:10.1002/9780470504437. ISBN 9780470504437

- Alt, H. G.; Koppl, A. (2000). "Effect of the Nature of Metallocene Complexes of Group IV Metals on Their Performance in Catalytic Ethylene and Propylene Polymerization". Chem. Rev. 100 (4): 1205–1222. doi:10.1021/cr9804700. PMID 11749264

- Caseri, Walter (2014). "Initial Organotin Chemistry". Journal of Organometallic Chemistry. 751: 20–24. doi:10.1016/j.jorganchem.2013.08.009

- "Polypropylene Production via Gas Phase Process, Technology Economics Program". Intratec. 2012. ISBN 978-0-615-66694-5

- Gielen, Marcel (1973). "From kinetics to the synthesis of chiral tetraorganotin compounds". Acc. Chem. Res. 6: 198–202. doi:10.1021/ar50066a0

- Sander H.L. Thoonen; Berth-Jan Deelman; Gerard van Koten (2004). "Synthetic aspects of tetraorganotins and organotin(IV) halides" (PDF). Journal of Organometallic Chemistry (689): 2145–2157

- Bochmann, M. (1994). Organometallics 1, Complexes with Transition Metal-Carbon σ-Bonds. New York: Oxford University Press. pp. 69–71. ISBN 9780198558132

- G. G. Graf "Tin, Tin Alloys, and Tin Compounds" in Ullmann's Encyclopedia of Industrial Chemistry, 2005 Wiley-VCH, Weinheim doi:10.1002/14356007.a27_049

- S. Gómez-Ruiz; et al. (2008). "Study of the cytotoxic activity of di and triphenyltin(IV) carboxylate complexes". Journal of Inorganic Biochemistry. 102 (12): 2087–96. doi:10.1016/j.jinorgbio.2008.07.009. PMID 18760840

- Davies, Alwyn George. (2004) Organotin Chemistry, 2nd Edition Weinheim: Wiley-VCH. ISBN 978-3-527-31023-4

- Chandrasekhar, Vadapalli; Nagendran, Selvarajan; Baskar, Viswanathan (2002). "Organotin assemblies containing Sn/O bonds". Coordination Chemistry Reviews. 235: 1–52. doi:10.1016/S0010-8545(02)00178-9

4

Organoelements in Organometallic Chemistry

The topics discussed in the chapter are of great importance to broaden the existing knowledge on the compound in organometallic chemistry. The main factors affecting the reactions of organometallic compounds along with reaction patterns have been discussed in the following section. It has been carefully written to provide an easy understanding of the subject matter.

As(V) and Sb(V) Compound

Organic chemistry of non-metal phosphorus, metalloids such as arsine and antimony along with metallic element bismuth is termed as organoelement chemistry.

The importance given to organoarsenic compounds earlier due to their medicinal values was waded out after antibiotics were discovered and also their carcinogenic and toxic properties were revealed. Also, the synthetically important organometallic compounds of group 13 and 14 masked the growth of group 15 elements. However, the organoelement compounds of phosphorus, arsenic and antimony find usefulness as ligands in transition metal chemistry due to their σ-donor and π-acceptor abilities which can be readily tuned by simply changing the substituents. These donor properties are very useful in tuning them as ligands to make suitable metal complexes for metal mediated homogeneous catalysis. Although organoelement compounds can be formed in both +3 (trivalent and tricoordinated) and +5(pentavalent and tetra or pentacoordinated) oxidation states, trivalent compounds are important in coordination chemistry.

For organoelement compounds of group 15, the energy of E—C bond decreases in the order, E = P > As > Sb > Bi, and in the same sequence E—C bond polarity increases.

Organometallic Compounds of As(V) and Sb(V)

Due to the strong oxidizing nature of pentahalides, the direct alkylation or arylation to generate ER5 is not feasible, but can be prepared in two steps.

A few representative methods of preparation are given below:

$$Me_3As \xrightarrow{Cl_2} Me_3AsCl_2 \xrightarrow[Et_2O]{MeLi} Me_5As$$

$$Ph_3Sb + PhI \rightarrow Ph_4SbI \xrightarrow[-LiI]{PhLi} Ph_5Sb$$

$$Ph_3Bi \xrightarrow[-SO_2]{SO_2Cl_2} Ph_3BiCl_2 \xrightarrow[-MgXCl]{PhMgX} Ph_5Bi$$

Structures and Properties

Pentaalkyl or pentaaryl derivatives are moderately thermally stable. On heating above 100°C, they form trivalent compounds as shown below:

$$Me_5As \xrightarrow{T > 100^0} Me_3As + CH_4 + CH_2CH_2$$

$$Ph_5Sb \xrightarrow{T > 200^0} Ph_3Sb + Ph - Ph$$

Reaction with water,

$$Me_5As + H_2O \rightarrow Me_4AsOH + MeH$$

Pentavalent compounds readily form "tetrahedral onium" cations and "octahedral and hexacoordinatged ate" anions.

$$Ph_5E + BPh_3 \rightarrow [Ph_4E][BPh_4] \quad (E = As, Sb, Bi)$$

$$Ph_5E + LiPh \rightarrow Li[EPh_6]$$

In solid state, Ph_5As adopts trigonal bipyramidal geometry, whereas Ph_5Sb prefers square based pyramidal geometry although the energy difference between the two is marginal.

The salts of the type $[R_4E]^+$ adopt tetrahedral geometry, whereas hexacoordinated anions $[R_6E]^-$ assume octahedral geometry.

Mixed organo-halo compounds of the type RnEX5-n adopt often dimeric structures due to the presence of lone pairs of electrons on X which can readily coordinate to the second molecule. The following structural types can be anticipated.

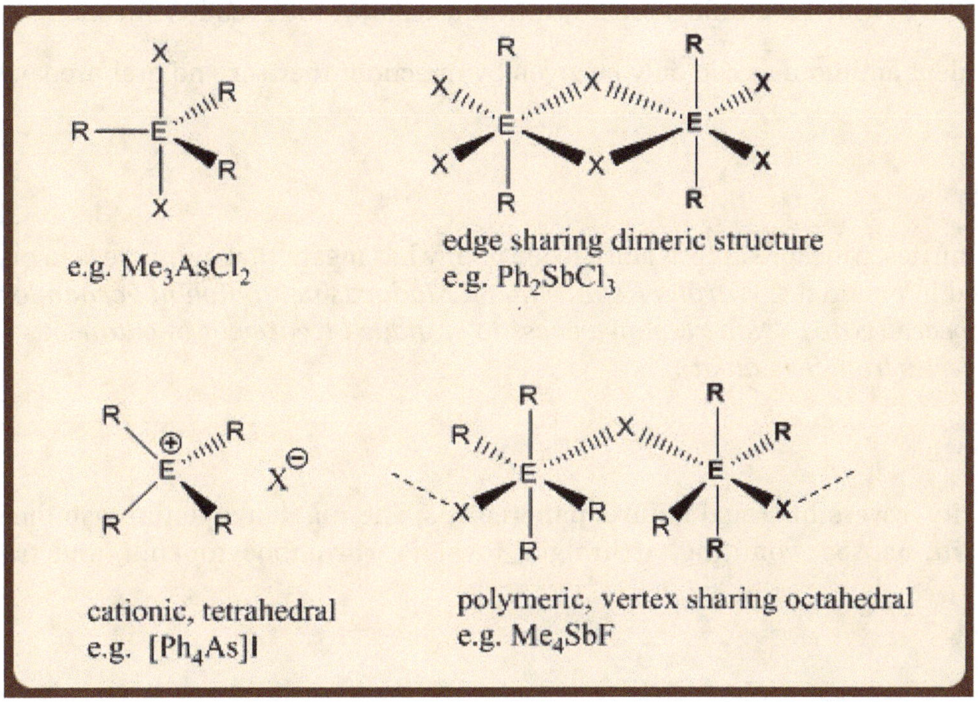

e.g. Me_3AsCl_2

edge sharing dimeric structure
e.g. Ph_2SbCl_3

cationic, tetrahedral
e.g. $[Ph_4As]I$

polymeric, vertex sharing octahedral
e.g. Me_4SbF

The thermal stability of R_nEX_{5-n} decreases with decreasing 'n'. Thermal reactions are essentially the reverse reactions of addition reactions used in the preparation of R_5E.

$$R_3SbX_2 \xrightarrow{\Delta T} R_2SbX + RX$$

$$Ph_3AsCl_2 \xrightarrow[100^\circ C]{CO_2} Ph_2AsCl + Cl_2$$

$$Me_2AsCl_3 \xrightarrow{50^\circ C} MeAsCl_2 + MeCl$$

Pentavalent Antimonial

Pentavalent antimonials (also abbreviated pentavalent Sb or Sb^V) are a group of compounds used for the treatment of leishmaniasis. They are also called pentavalent antimony compounds.

Types

The first pentavalent antimonial used was urea stibamine: first introduced in the 1930s, it fell out of favour in the 1950s due to higher toxicity compared to sodium stibogluconate.

The compounds currently available for clinical use are:

- sodium stibogluconate (*Pentostam*; manufactured by GlaxoSmithKline; available in United States [through the Centers for Disease Control only] and UK), which is administered by slow intravenous injection.

- meglumine antimoniate (*Glucantim*; manufactured by Aventis; available in Brazil, France and Italy), which is administered by intramuscular or intravenous injection.

The pentavalent antimonials can only be given by injection: there are no oral preparations available.

Alternatives

In many countries, widespread resistance to antimony has meant that amphotericin or miltefosine are now used in preference.*Wardha Refai, Nayani Madarasingha, Rohini Fernandopulle, Nadira Karunaweera (2016). "Non-responsiveness to standard treatment in cutaneous leishmaniasis: A case series from Sri Lanka".*

Side Effects

Cardiotoxicity, reversible renal failure, pancreatitis, anemia, leukopenia, rash, headache, abdominal pain, nausea, vomiting, arthralgia, myalgia, thrombocytopenia, and transaminase elevation.

As(III) and Sb(III) Compound

Direct Synthesis

Mono- Derivatives

$$2As + 3MeBr \xrightarrow[Cu]{\Delta T} Me_2AsBr + MeAsBr_2$$

$$Me_2AsBr + PhLi \rightarrow Me_2AsPh$$

$$MeAsBr_2 + 2PhLi \rightarrow MeAsPh_2$$

$$EX_3 + 3RMgX \rightarrow R_3E + 3MgX_2$$

$$EX_3 + 3RLi \rightarrow R_3E + 3LiX$$

Bis Derivatives

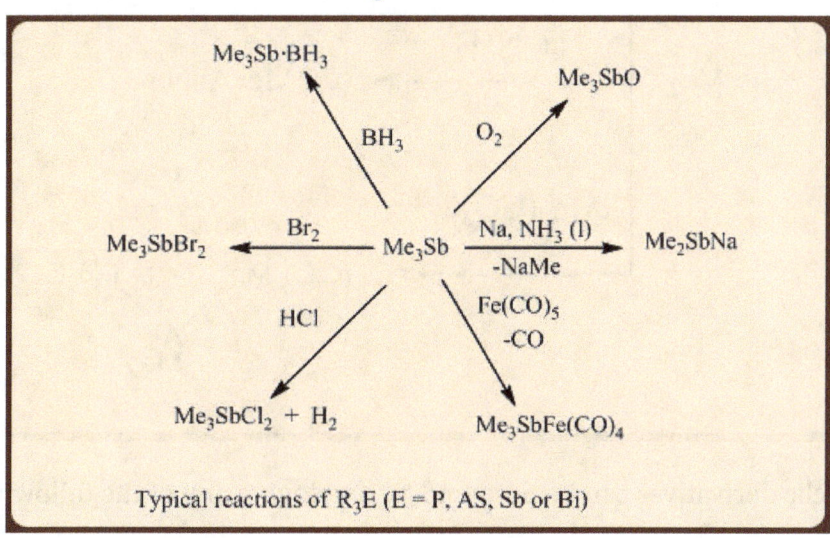

In a similar way, a variety of bisphosphines and arsines can be generated.

Reactions of Trialkyl Derivatives, R₃E

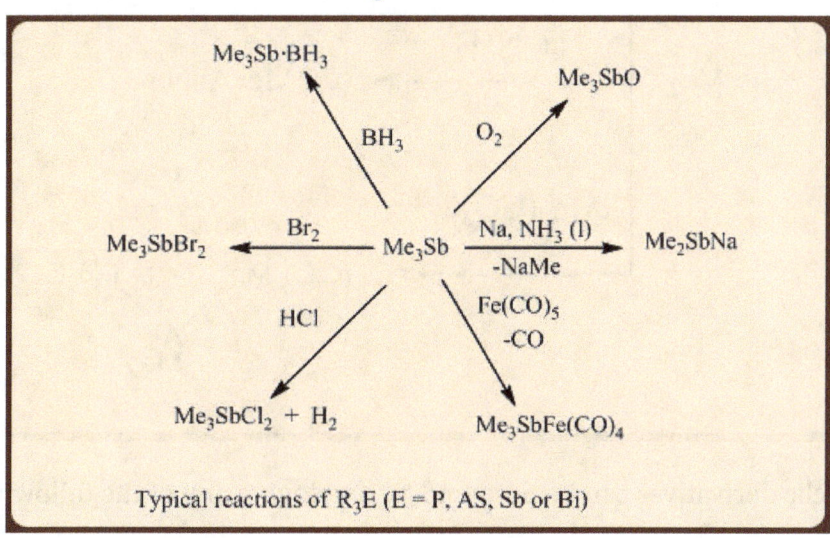

Typical reactions of R₃E (E = P, AS, Sb or Bi)

The transition metal chemistry of R_3E, phosphines, arsines or stibines has been extensively studied because of their distinct donor and acceptor properties. Among them, the phosphines or tertiary phosphines (R_3P) are the most valuable ligands in metal mediated homogeneous catalysis. Interestingly, the steric and electronic properties can be readily tuned by changing the substituents on phosphorus atoms.

Properties

Trialkyl derivatives are highly air-sensitive liquids with low boiling points and some of them are even pyrophyric. Triphenyl derivatives are solids at room temperature and are moderately stable and oxidizing agents such as $KMnO_4$, H_2O_2 or TMNO are needed for oxidation to form $Ph_3E=O$.

Cyclic and acyclic derivatives containing E—E bonds

E—E Single Bonds

The E—E bond energies suggest that they do not have greater stability and the stability decreases down the group.

The simplest molecules include $Ph_2P—PPh_2$, $Me_2As—AsMe_2$ prepared by coupling reactions:

$$Me_2AsH + Me_2AsCl \xrightarrow{-HCL} Me_2As-AsMe_2$$
$$2Ph_2BiCl \xrightarrow{Na,\ NH_3} Ph_2Bi-BiPh_2$$

The weakness of E—E bonds accounts for many interesting reactions and a few of such reactions are listed below:

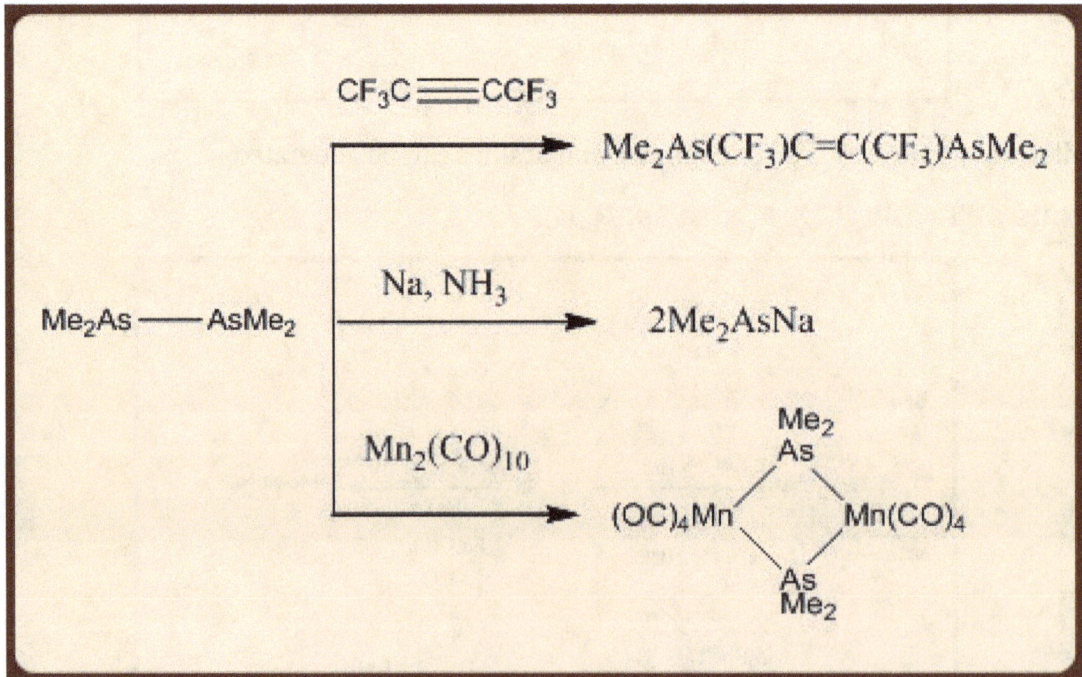

Cyclic and polycyclic derivatives can be prepared by employing any of the following methods:

$$8RAsCl_2 + 4AsCl_3 + 14 Mg \xrightarrow[-14MgCl_2]{}$$

(R = ^{t}Bu)

Confirm that the octahedral structure of $[Ph_6Bi]^-$ is consistent with VSEPR theory.

Octahedral similar to PF_6^-

5 (Bi valence electrons) + 6 (each Ph) + 1 (-ve charge) = 12 electrons

i.e. six pairs, octahedral geometry

Phosphines

Classification of Ligands by Donor Atoms

Ligand is a molecule or an ion that has at least one electron pair that can be donated. Ligands may also be called Lewis bases; in terms of organic chemistry, they are 'nucleophiles'.

Metal ions or molecules such as BF_3 (with incomplete valence electron shells (electron deficient) are called Lewis acids or electrophiles).

Why do molecules like H_2O or NH_3 give complexes with ions of both main group and transition metals. E.g $[Al(OH_2)_6]_3^+$ or $[Co(NH_3)_6]_3^+$

Why other molecules such as PF_3 or CO give complexes only with transition metals.

Although PF_3 or CO give neutral molecules such as $Ni(PF_3)_4$ or $Ni(CO)_4$ or $Cr(CO)_6$.

Why do, NH_3, amines, oxygen donors, and so on, not give complexes such as $Ni(NH_3)_4$.

Classical or Simple Donor Ligands

Act as electron pair donors to acceptor ions or molecules, and form complexes of all types of Lewis acids, metal ions or molecules.

Non-classical ligands, π-bonding or π-acid ligands: Form largely with transition metal atoms.

In this case special interaction occurs between the metals and ligands

These ligands act as both σ-donors and π-acceptors due to the availability of empty orbitals of suitable symmetry, and energies comparable with those of metal t_2g (non-bonding) orbitals.

e.g. Consider PR_3 and NH_3: Both can act as bases toward H^+, but P atom differs from N in that PR3 has σ* orbitals of low energy, whereas in N the lowest energy d orbitals or σ* orbitals are far too high on energy to use.

Consider CO that do not have measurable basicity to proton, yet readily reacts with metals like Ni that have high heats of atomization to give compounds like $Ni(CO)_4$.

Ligands may also be classified electronically depending upon how many electrons that they contribute to a central atom. Atoms or groups that can form a single covalent bond are one electron donors.

EXAMPLES: F, SH, CH_3 etc.,

Compounds with an electron pair are two-electron donors

EXAMPLE: NH_3, H_2O, PR_3 etc.,

Bonding in Metal –Carbonyl and Metal-Phosphines

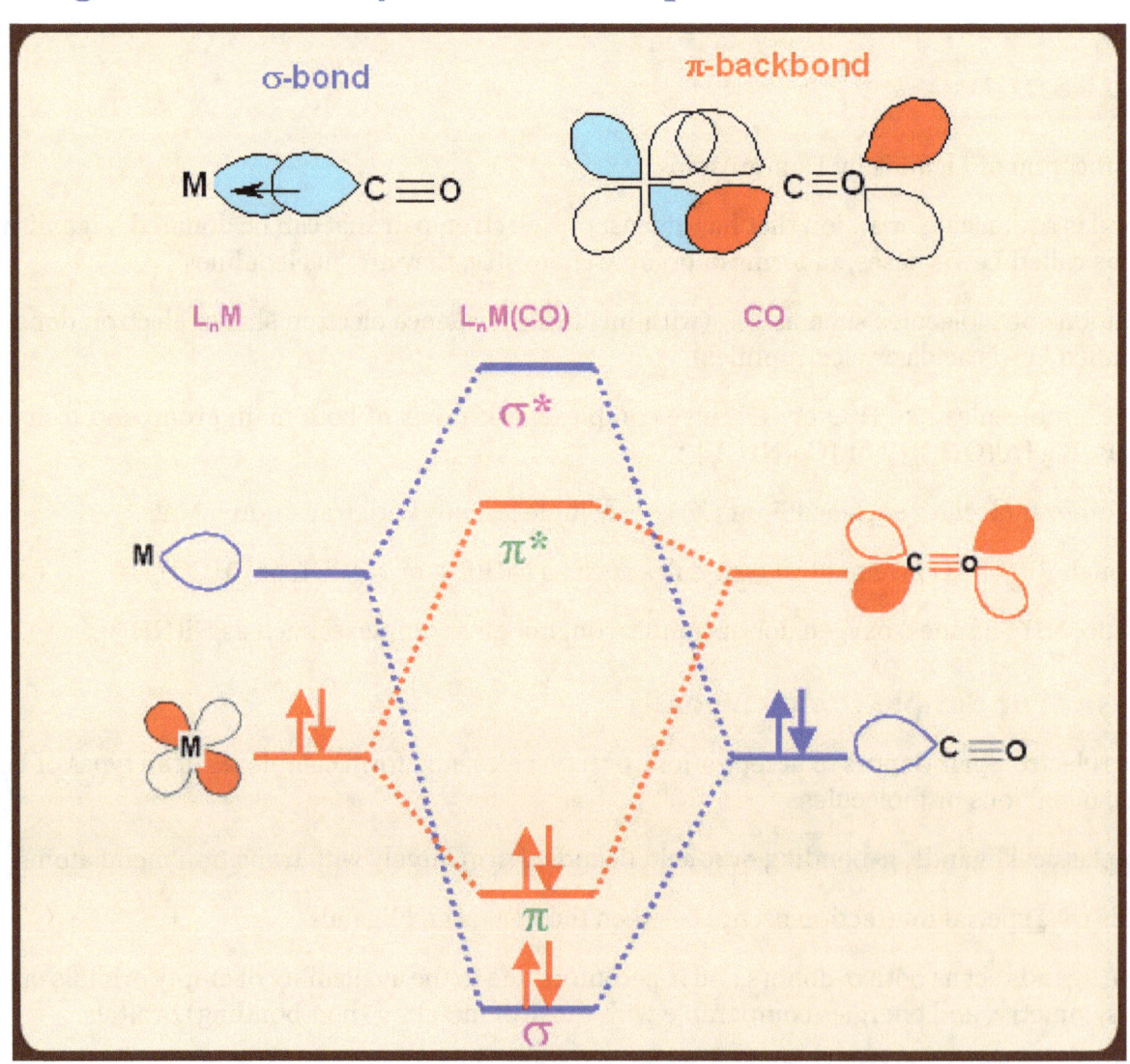

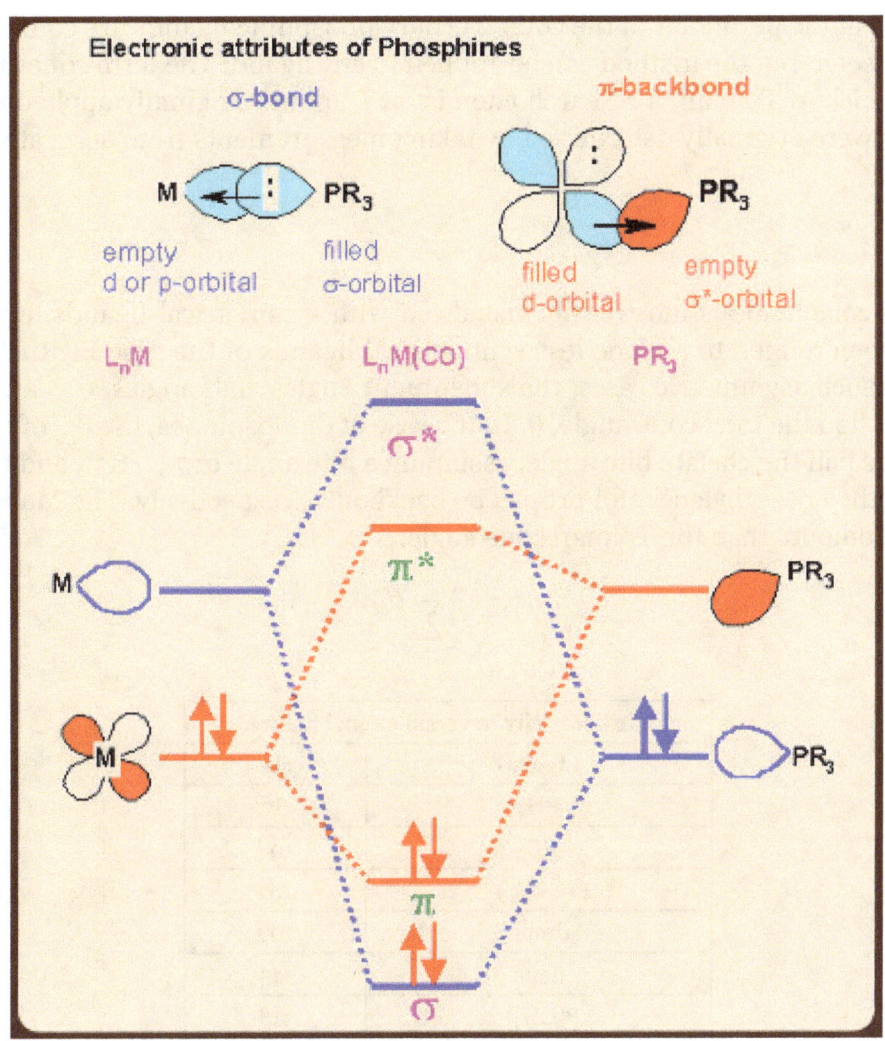

Ligand Cone Angle

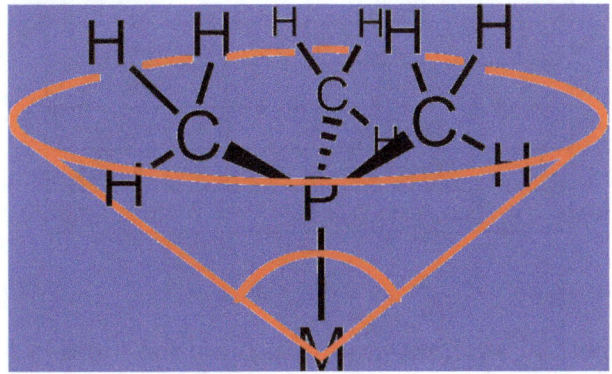

The ligand cone angle (a common example being the Tolman cone angle or θ) is a measure of the size of a ligand. It is defined as the solid angle formed with the metal at the vertex and the hydrogen atoms at the perimeter of the cone. Tertiary phosphine ligands are commonly classified using this parameter, but the method can be applied to any ligand. The term cone angle was introduced by Chadwick A. Tolman, a research chemist at Dupont. Originally applied to phosphines, the cone angles were originally determined by taking measurements from accurate physical models of them.

Asymmetric Cases

The concept of cone angle is most easily visualized with symmetrical ligands, e.g. PR_3. But the approach has been refined to include less symmetrical ligands of the type PRR'R'' as well as diphosphines. In such asymmetric cases, the substituent angles' half angles, $\theta_{i/2}$, are averaged and then doubled to find the total cone angle, θ. In the case of diphosphines, the $\theta_{i/2}$ of the backbone is approximated as half the chelate bite angle, assuming a bite angle of 74°, 85°, and 90° for diphosphines with methylene, ethylene, and propylene backbones, respectively. The Manz cone angle is often easier to compute than the Tolman cone angle:

$$\theta = \frac{2}{3} \sum_i \frac{\theta_i}{2}$$

Cardiotoxicity, reversible renal failure,	
Ligand	**Angle (°)**
PH_3	87
PF_3	104
$P(OCH_3)_3$	107
dmpe	107
depe	115
$P(CH_3)_3$	118
dppm	121
dppe	125
dppp	127
$P(CH_2CH_3)_3$	132
dcpe	142
$P(C_6H_5)_3$	145
$P(cyclo\text{-}C_6H_{11})_3$	179
$P(t\text{-}Bu)_3$	182
$P(C_6F_5)_3$	184
$P(2,4,6\text{-}Me_3C_6H_2)_3$	212

Variations

The Tolman cone angle method assumes empirical bond data and defines the perimeter as the maximum possible circumscription of an idealized free-spinning substituent. In contrast, the solid-angle concept derives both bond length and the perimeter from empirical solid state crystal structures. There are advantages to each system.

If the geometry of a ligand is known, either through crystallography or computations, an exact cone angle (θ) can be calculated. No assumptions about the geometry are made, unlike the Tolman method.

Application

The concept of cone angle is of practical importance in homogeneous catalysis because the size of the ligand affects the reactivity of the attached metal center. In a famous example, the selectivity of hydroformylation catalysts is strongly influenced by the size of the coligands. Despite being monovalent, some phosphines are large enough to occupy more than half of the coordination sphere of a metal center.

Steric Factors in Phosphines (Tolman's Cone Angle)

Cone angle is very useful in assessing the steric properties of phosphines and their coordination behavior.

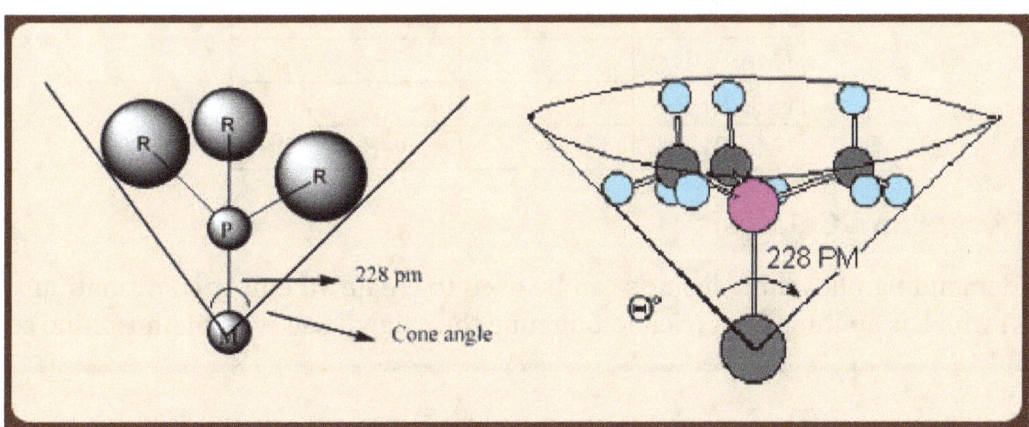

The electronic effect of phosphines can be assessed by IR and NMR spectroscopic data especially when carbonyls are co-ligands. In a metal complex containing both phosphines and carbonyl, the $v(CO)$ frequencies would reveal the σ-donor or π-acceptor abilities of phosphines. If the phosphines employed are strong σ-donors, then more electron density would move from M (t_2g orbitals)- $\pi^*(CO)$ and as a result, a lowering in the $v(CO)$ is observed. In contrast, if a given phosphine is a poor σ-donor but strong π-acceptor, then phosphine(σ^*-orbitals) also compete with CO for back bonding which results in less lowering in $v(CO)$ frequency.

Another important aspect is the steric size of PR3 ligands, unlike in the case of carbonyls, which can be readily tuned by changing R group. This is of great advantage in transition metal chemistry, especially in metal mediated catalysis, where stabilizing the metals in low coordination states is very important besides low oxidation states. This condition can promote oxidative addition at the metal centre which is an important step in homogeneous catalysis. The steric effects of phosphines can be quantified with Tolman's cone angle.

Cone angle can be defined as a solid angle at metal at a M—P distance of 228 pm which encloses the van der Waal's surfaces of all ligand atoms or substituents over all rotational orientations. The cone angles for most commonly used phosphines are listed in the following table.

Phosphine	Cone Angle (°)
PH_3	87
PF_3	104
$P(OMe)_3$	107
PMe_3	118
PMe_2Ph	122
PEt_3	132
PPh_3	145
PCy_3	170
$P(Bu^t)_3$	182
$P(mesityl)_3$	212

Phosphines with different cone angles versus coordination number for group 8 metals:

ML_4	ML_3	ML_2
$(Me_3P)_4Ni$		
$(Me_3P)_4Pd$		
$(Me_3P)_4Pt$	$(Ph_3P)_3Pt$	$(tert\text{-}Bu_3P)_2Pt$

Tolman Angle and Catalysis

Sterically demanding phosphine ligands can be used to create an empty coordination site (16 VE complexes) which is an important trick to fine tune the catalytic activity of phosphine complexes.

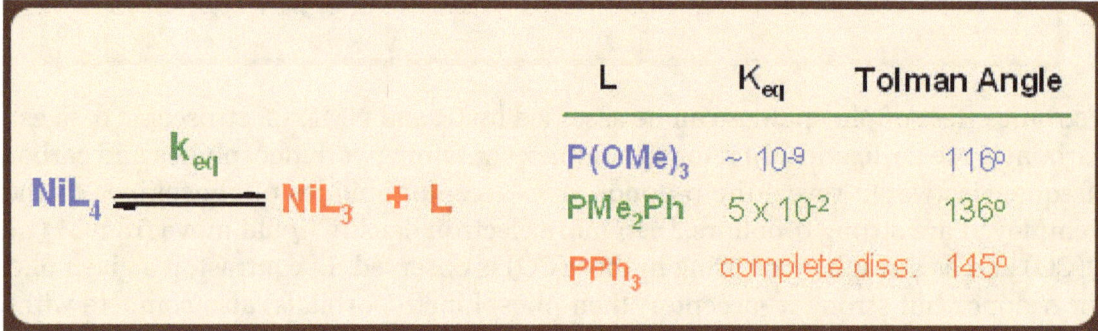

Zinc and Cadmium

Dialkyl compounds of Zn, Cd and Hg do not associate through alkyl bridges. Dialkylzinc compounds are only weak Lewis acids, organocadmium compounds are even weaker, and organomercury compounds do not act as Lewis acids except under special circumstances.

The Group 12 metals form linear molecular compounds, such as $ZnMe_2$, $CdMe_2$ and $HgMe_2$, that are not associated in solid, liquid or gaseous state or in hydrocarbon solution.

They form 2c, 2e bonds. Unlike Be and Mg analogs, they do not complete their valence shells by association through alkyl bridges. The bonding in these molecules are similar to d^{10} metals such as Cu^I, Ag^I and Au^I with linear geometry ($[N\equiv C\text{-}M\text{-}C\equiv N]^-$, M = Ag or Au). This tendency is sometimes rationalized by invoking pd hybridization in the M^+ ion, which leads to orbitals that favor linear attachment of ligands (similar to spd hybridization).

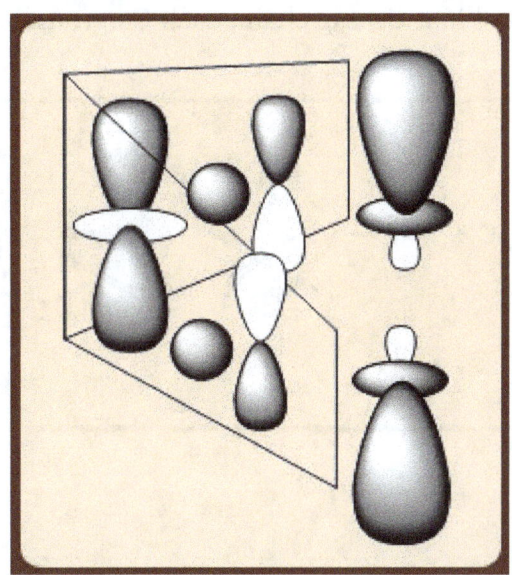

The preference for the linear coordination may be due to the similarity inenergy of the outer ns, np and (n-1)d orbitals, which permits the formation of collinear spd hybrids.

The hybridization of s, p_z and d_z2 with the choice of phases shown here produces a pair of collinear orbitals that can be used to form strong σ-bonds.

Organozinc and Organocadmium Compounds

Convenient route is metathesis with alkylaluminium or alkyllithium compounds.

With alkyllithium compounds it is the electronegativity which is decisive, whereas between Al and Zn it is hardness considerations correctly predict the formation of softer ZnCH3 and harder AlCl pairs.

$$ZnCl_2 + Al_2Me_6 \rightarrow ZnMe_2 + Al_2Cl_2Me_4$$

Alkylzinc compounds are pyrophoric and readily hydrolyzed, whereas alkylcadmium compounds react more slowly with air. Due to mild Lewis acidity, dialkylzinc and dialkylcadimum compounds form stable complexes with amines, especially with chelating amines.

The Zn—C has greater carbanionic character than the Cd—C bond.

For example, addition of alkylzinc compounds across the carbonyl group of a ketone:

$$ZnMe_2 + (CH_3)_2 C = O \rightarrow (CH_3)_2 C - O - ZnCH_3$$

This reaction do not proceed with the less polar alkylcadmium or alkylmercury compounds, but organolithium, organomagnesium and organoaluminium compounds can promote this reaction readily since all of which contain metals with lower electronegativity than zinc.

Interestingly, the cyclodipentadienyl compounds are structurally unusual. CpZnMe is monomeric in the gas phase with a pentahapto Cp group.

In the solid state it is associated in a zig-zag chain, each Cp group being pentahapto with respect to two Zn atoms.

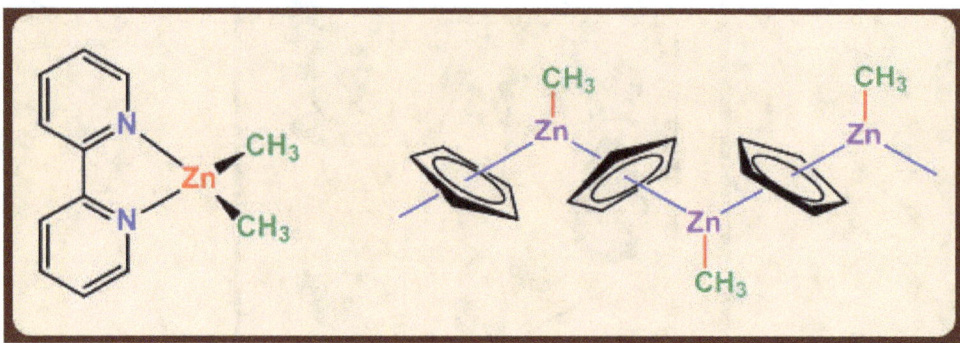

Mercury

$$2RMgX + HgX_2 \rightarrow HgR_2 + MgX_2$$

Reaction proceeds due to both electronegativity and hardness considerations.

Dialkylmercury compounds are very versatile starting materials for the synthesis of many organometallic compounds of more elctropositive metals by transmetallation.

However, owing to high toxicity of alkylmercury compounds, other synthons are preferred. In striking contrast to the high sensitivity of dimethylzinc to oxygen, dimethylmercury survives exposure to air.

Mercury Toxicity

The toxicity of mercury arises from the very high affinity of the soft Hg atom for sulfhydryl (—SH) groups in enzymes. Simple mercury-sulfur compounds have been studied as potential analogs of natural systems. The Hg atoms are most commonly four-coordinated, as in $[Hg_2(SMe)_6]^{2-}$.

Mercury poisoning was a serious concern even from early days. Issac Newton, Alfred Stock worked in the early 20th century. Later in 60s awareness came following the incidence of brain damage and death it caused among the inhabitants in Minamata, Japan. Mercury from a plastic company was allowed to escape into a bay where it found its way into fish that were later eaten. Research has shown that bacteria found in sediments are capable of methylating mercury, and that species such $HgMe_2$ and $[HgCH_3]^+$ enter the food chain because they readily penetrate cell walls. The bacteria appear to produce $HgMe_2$ as a means of eliminating toxic mercury ions through their cell walls and into the environment.

References

- Immirzi, A.; Musco, A. (1977). "A method to measure the size of phosphorus ligands in coordination complexes". Inorg. Chim. Act. 25: L41–L42. doi:10.1016/S0020-1693(00)95635-4. Retrieved 2009-12-04

- Tolman, Chadwick A. (1970-05-01). "Phosphorus ligand exchange equilibriums on zerovalent nickel. Dominant role for steric effects". J. Am. Chem. Soc. 92 (10): 2956–2965. doi:10.1021/ja00713a007

- Tolman, C. A.; Seidel, W. C.; Gosser, L. W. (1974-01-01). "Formation of three-coordinate nickel(0) complexes by phosphorus ligand dissociation from NiL_4". J. Am. Chem. Soc. 96 (1): 53–60. doi:10.1021/ja00808a009

- Tolman, C. A. (1977). "Steric Effects of Phosphorus Ligands in Organometallic Chemistry and Homogeneous Catalysis". Chem. Rev. 77 (3): 313–48. doi:10.1021/cr60307a002

- Niksch, Tobias; Görls, Helmar; Weigand, Wolfgang (2009). "The Extension of the Solid-Angle Concept to Bidentate Ligands". Eur. J. Inorg. Chem. 2010 (1): 95–105. doi:10.1002/ejic.200900825

- Bilbrey, Jenna A.; Kazez, Arianna H.; Locklin, J.; Allen, Wesley D. (2013). "Exact ligand cone angles". J. Comput. Chem

- Evans, D.; Osborn, J. A.; Wilkinson, G. (1968). "Hydroformylation of Alkenes by Use of Rhodium Complex Catalyst". J. Chem. Soc. 33 (21): 3133–3142. doi:10.1039/J19680003133

5

An Integrated Study of Metal Alkyls and Hydrides

Metal alkyls are readily available on Earth and are a major source of stabilized carbanions. These alkyls can be divided into two categories – stable alkyls and agostic alkyls. The section also explores hydrides and its facets. Hydrides have assumed prominence in organometallic chemistry due to its ability to undergo insertion reaction and form various organometallic compounds. The nature of hydrogen in hydrides can vary from being protic to hydidic. This section has been carefully written to provide an easy understanding of the varied facets of alkyls and hydrides.

Metal Alkyl

Transition metal σ−bonded organometallic compounds like the metal alkyls, aryls and the hydrides derivatives are by for the most common organometallic species encountered in the world of chemistry. Yet, these compounds remained elusive till as late as the 1960s and the 1970s.

Historical Background

Metal alkyls of the main group elements namely, Li, Mg, Zn, As and Al, have been known for a long time and which over the years have conveniently found applications in organic synthesis whereas development on similar scale and scope in case of the transition metal counterparts were missing till only recently. The origin of the organometallic compounds traces back to 1757, when Cadet prepared a foul smelling compound called cacodyl oxide from As_2O_3 and CH_3COOK, while working in a military pharmacy in Paris. Years later in 1840, R. W. Bunsen gave the formulation of cacodyl oxide as $Me_2As-O-AsMe_2$. The next known transition metal organometallic compound happens to be Et_2Zn, which was prepared serendipitously in 1848 from the reaction of ethyl iodide (EtI) and Zn with the objective of generating free ethyl radical. Frankland further synthesized alkyl mercury halides like, CH_3HgI, from the reaction of methyl iodide (CH_3I) and Hg in sunlight. It is important to note that the dialkyl mercury, R_2Hg, and the dialkyl zinc, R_2Zn, have found applications as alkyl transfer reagents in the synthesis of numerous main group organometallic compounds.

Another notable development of the time was of the preparation of Et_4Pb from ethyl iodide (EtI) and Na/Pb alloy by C. J. Lowig and M. E. Schweizer in 1852. They subsequently extended the same method for the preparation of the Et_3Sb and Et_3Bi compounds. In 1859, aluminumalkyliodides, R_2AlI, were prepared by W. Hallwachs and A Schafarik from alkyl iodide (RI) and Al. The year 1863 saw the preparation of organochlorosilanes, $RmSiCl_{4-m}$, by C. Friedel and J. M. Crafts while the year 1866 saw the synthesis of halide-free alkyl magnesium compound, Et_2Mg, by J. A. Wanklyn from the reaction of Et_2Hg and Mg. In 1868, M. P. Schutzenberger reported the first metal–carbonyl complex in the form of $[Pt(CO)Cl_2]_2$. In 1890, the first binary metal–carbonyl compound,

$Ni(CO)_4$ was reported by L. Mond, who later founded the well–known chemical company called ICI (Imperial Chemical Industries). In 1909, W. J. Pope reported the first σ–organotransition metal compound in the form of $(CH_3)_3PtI$. In 1917, the alkyllithium, RLi, compounds were prepared by W. Schlenk by transalkylation reactions. In 1922, T. Midgley and T. A. Boyd reported the utility of Et_4Pb as an antiknock agent in gasoline. A. Job and A. Cassal prepared $Cr(CO)_6$ in 1927. In 1930, K. Ziegler showed the utility of organolithium compounds as alkylating agent while in the following year in 1931, W. Heiber prepared $Fe(CO)_4H_2$ as the first transition metal–hydride complex. O. Roelen discovered the much renowned hydroformylation reaction in 1938, that went on to become a very successful industrial process worldwide.

The large scale production and the use of silicones were triggered by E. G. Rochow, when he reported the 'direct synthesis' from methyl chloride (CH_3Cl) and Si using Cu catalyst at 300 °C in 1943. The landmark compound, ferrocene $(C_5H_5)_2Fe$, known as the first sandwich complex was obtained by P. Pauson and S. A. Miller in 1951. H. Gilman introduced the important utility of organocuprates when he prepared $LiCu(CH_3)_2$, in 1952. In the subsequent year 1953, G. Wittig found a new method of synthesizing olefins from phosphonium ylides and carbonyl compounds that fetched him a Nobel prize in 1979. The year 1955 turned out to be a year of path breaking discoveries with E. O. Fischer reporting the rational synthesis of bis(benzene)chromium, $(C_6H_6)_2Cr$ while K. Ziegler and G. Natta announcing the ground breaking polyolefin polymerization process that subsequently gave them the Nobel prizes, E. O. Fischer sharing with G. Wilkinson in 1973 while K. Ziegler and G. Natta shared the same in 1963. In 1956, H. C. Brown reported hydroboration for which he too received the Nobel prize in 1979. In 1963, L. Vaska reported the famous Vaska's complex, $trans–(PPh_3)_2Ir(CO)Cl$, that reversibly binds to molecular oxygen. In 1964, E. O. Fischer reported the first carbene complex, $(CO)_5WC(OMe)Me$. In 1965, G. Wilkinson and R. S. Coffey reported the Wilkinson catalyst, $(PPh_3)_3RhCl$, for the hydrogenation of alkenes. In 1973, E. O. Fischer synthesized the first carbyne complex, $I(CO)_4Cr(CR)$.

After the early 1970s, there were tremendous outburst in activity, in the area of transition metal organometallic chemistry leading to phenomenal developments having far-reaching consequences in various branches of the main stream and interfacial chemistry. Several Nobel prizes that have been awarded to the area in recent times fully recognized the significance of these efforts with Y. Chauvin, R. R. Schrock, and R. H. Grubbs winning it in 2005 for olefin metathesis and Akira Suzuki, Richard F Heck and E. Negishi receiving the same for the Pd catalyzed C–C cross-coupling reactions in organic synthesis in 2010.

Metal Alkyls

In day to day organic synthesis, particularly from the application point of view, the metal alkyls are often perceived as a source of stabilized carbanions for reactions with various electrophiles. The extent of stabilization of alkyl carbanions in metal alkyl complexes depend upon the nature of the metal cations. For example, the alkyls of electropositive metals like that of Group 1 and 2, Al and Zn are regarded as polar organometallics as the alkyl carbanions remain weakly stabilized while retaining strong nucleophilic and basic character of a free anion. These polar alkyls are extremely air and moisture sensitive as in their presence they often get hydrolyzed and oxidized readily. Similar high reactivity was also observed in case of the early transition metal organometallic compounds particularly of Ti and Zr. On the contrary the late transition metal organometallic compounds are much less reactive and stable. For example, the Hg–C bond of $(Me–Hg)^+$ cation is indefinitely stable in aqueous H_2SO_4 solution in air. Thus, on moving from extremely ionic Na alkyls to highly polar covalent Li and Mg alkyls and to essentially covalent late–transition metal alkyls, a steady decrease in reactivity is observed. This trend can be correlated to the stability of alkyl carbanions that also depended on the nature of hybridization of the carbon center, with sp^3 hybridized carbanions being the least stable and hence most reactive, followed by the sp^2 carbanions being moderately stable while the sp carbanions being the least reactive and most stable. The trend also correlates well with the respective pKa values observed for CH_4 (pKa = ~50), C_6H_6 (pKa = ~43) and $RC{\equiv}CH$ (pKa = ~25).

Stable Alkyls

As has been mentioned earlier, that the β–elimination is a crucial destabilizing influence on the transition metal organometallic complexes. Hence, inhibition of this decomposition pathway leads to increased stabilities of organometallic compounds. Thus, many stable alkyl transition metal complexes do not possesses β–hydrogens like, $W(Me)_6$ and $Ti(CH_2Ph)_4$. In some cases despite the presence of the β–hydrogens the organometallic complexes are stable as the β–hydrogens are deposed away from the metal center like in, $Cr(CHMe_2)_4$, and $Cr(CMe_3)_4$. In this category of stable transition metal organometallic compounds also falls the ones that contain β–hydrogens but cannot β–eliminate owing to the formation of a olefinic bond at a bridgehead, which is unfavorable, like in $Ti(6-norbornyl)_4$ and $Cr(1-adamantyl)_4$. Lastly, some 18 VE metal complexes are stable, again despite having β–hydrogens, for reasons of being electronically as well as coordinatively saturated at the metal center owing to attaining the stable 18 electron configuration.

Agostic Alkyls

Agostic alkyls are extremely rare but very interesting species that represents a frozen point in a β–elimination pathway that have fallen short of the completion of the decomposition reaction. Thus, these agostic alkyl complexes can be viewed as snap shots of a β–elimination trajectory thereby providing valuable mechanistic understanding of the decomposition reaction. The agostic interaction has characteristic signatures in various spectroscopic techniques as observed from the decreasing JC–H coupling constant values in the ^1H NMR and the ^{13}C NMR spectra and the lowering of the ^vC–H stretching frequencies in the IR spectroscopy. The agostic alkyl complexes can be definitively proven by X–ray diffraction or neutron diffraction studies. The agostic alkyls thus have activated C–H bonds which are of interest for their utility in chemical catalysis. Quite interestingly, many d^0 Ti agostic alkyl complexes do not β–eliminate primarily for the metal center being too electron deficient to donate electron to the σ* C–H orbital as required for the subsequent β–elimination process.

Reductive Elimination

Reductive elimination represents a major decomposition pathway of the metal alkyls. Opposite of oxidative addition, the reductive elimination is accompanied by the decrease in the oxidation state and the valence electron count of the metal by two units. The metal alkyl complexes may thus reductively eliminate with an adjacent hydrogen atom to yield an alkane, (R–H) or undergo the same with an adjacent alkyl group to give an even larger alkane (R–R) as shown below.

$$L_nMRH \;\rightarrow\; R-H+L_nM$$
$$L_nMR_2 \;\rightarrow\; R-R+L_nM$$

The reductive elimination is often facilitated by an electron deficient metal center and by sterically demanding ligand systems. Often d^8 metals like Ni(II), Pd(II), and Au(III) and d^6 metals in high oxidation state like, Pt(IV), Pd(IV), Ir(III), and Rh(III) exhibit reductive elimination.

Oxidative Addition

Unlike the reductive elimination that represents a decomposition pathway of metal alkyls, the oxidative addition reaction represents a useful method for the formation of the metal alkyl complexes. The oxidative addition thus leads to increase in valence electron count and the oxidation state of the metal center by two units. The oxidative addition reactions are often facilitated by low valent electron rich metal centers and by less sterically demanding ligands.

$$L_nM + A—B \longrightarrow L_nM\overset{A}{\underset{B}{\diagup}}$$

$$\text{16VE} \qquad\qquad\qquad\qquad \text{18VE}$$

Halide Elimination

β–halide elimination is observed for the early transition metals and the f–block elements resulting in the formation of stable alkyl halides. The phenomenon is mostly seen in case of the metal fluorides and arise owing to the very high alkyl–fluoride bond strengths that favor the halide elimination.

Metal Hydrides

Metal hydrides occupy an important place in transition metal organometallic chemistry as the M–H bonds can undergo insertion reactions with a variety of unsaturated organic substrates yielding numerous organometallic compounds with M–C bonds. Not only the metal hydrides are needed as synthetic reagents for preparing the transition metal organometallic compounds but they also are required for important hydride insertion steps in many catalytic processes. The first transition metal hydride compound was reported by W. Heiber in 1931 when he synthesized $Fe(CO)_4H_2$. Though he claimed that the $Fe(CO)_4H_2$ contained Fe–H bond, it was not accepted until 1950s, when the concept of normal covalent M–H bond was widely recognized.

The metal hydride moieties are easily detectable in 1H NMR as they appear high field of TMS in the region between 0 to 60 ppm, where no other resonances appear. The hydride moieties usually couple with metal centers possessing nuclear spins. Similarly, the hydride moieties also couple with the adjacent metal bound phosphine ligands, if at all present in the complex, exhibiting characteristic cis (J = 15 – 30 Hz) and trans (J = 90 – 150 Hz) coupling constants. In the IR spectroscopy, the M–H frequencies appear between (1500 – 2200) cm–1 but their intensities are mostly weak. Crystallographic detection of metal hydride moiety is difficult as hydrogen atoms in general are poor scatterer of X–rays. Located adjacent to a metal atom in a M–H bond, the detection of hydrogen atom thus becomes challenging and as a consequence the X–ray crystallographic method systematically underestimates the M−H internuclear distance by ~ 0.1 Å. However, better data could be obtained by performing the X−ray diffraction studies at a low temperature in which the thermal motion of the atoms are significantly reduced. In light of these facts, the neutron diffraction becomes a powerful method for detection of the metal hydride moieties as hydrogen scatters neutrons more effectively and hence the M−H bond distances can be measured more accurately. A limitation of neutron diffraction method is that large sized crystals are required for the study.

Hydride

In chemistry, a hydride is the anion of hydrogen, H^-, or, more commonly, it is a compound in which one or more hydrogen centres have nucleophilic, reducing, or basic properties. In compounds that are regarded as hydrides, the hydrogen atom is bonded to a more electropositive element or group. Compounds containing hydrogen bonded to metals or metalloid may also be referred to as hydrides, even though in this case the hydrogen atoms can have a protic[clarification needed] character. Almost all of the elements form binary compounds with hydrogen, the exceptions being He, Ne, Ar, Kr, Pm, Os, Ir, Rn, Fr, and Ra.

Bonds

Bonds between hydrogen and other elements range from highly to somewhat covalent. Some hydrides, e.g. boron hydrides, do not conform to classical electron-counting rules, and the bonding is described in terms of multi-centered bonds, whereas the interstitial hydrides often involve metallic bonding. Hydrides can be discrete molecules, oligomers or polymers, ionic solids, chemisorbed monolayers, bulk metals (interstitial), and other materials. While hydrides traditionally react as Lewis bases or reducing agents, some metal hydrides behave as hydrogen-atom donors and act as acids.

Applications

- Hydrides such as sodium borohydride, lithium aluminium hydride, diisobutylaluminium hydride (DIBAL) and super hydride, are commonly used as reducing agents in chemical synthesis. The hydride adds to an electrophilic center, typically unsaturated carbon.

- Hydrides such as sodium hydride and potassium hydride are used as strong bases in organic synthesis. The hydride reacts with the weak Bronsted acid releasing H_2.

- Hydrides such as calcium hydride are used as desiccants, i.e. drying agents, to remove trace water from organic solvents. The hydride reacts with water forming hydrogen and hydroxide salt. The dry solvent can then be distilled or vac transferred from the "solvent pot".

- Hydrides are important in storage battery technologies such as nickel-metal hydride battery. Various metal hydrides have been examined for use as a means of hydrogen storage for fuel cell-powered electric cars and other purposed aspects of a hydrogen economy.

- Hydride complexes are catalysts and catalytic intermediates in a variety of homogeneous and heterogeneous catalytic cycles. Important examples include hydrogenation, hydro-formylation, hydrosilylation, hydrodesulfurization catalysts. Even certain enzymes, the hydrogenase, operate via hydride intermediates. The energy carrier Nicotinamide adenine dinucleotide reacts as a hydride donor or hydride equivalent.

Hydride Ion

Free hydride anions exist only under extreme conditions and are not invoked for homogeneous solution. Instead, many compounds have hydrogen centres with hydridic character.

Aside from electride, the hydride ion is the simplest possible anion, consisting of two electrons and a proton. Hydrogen has a relatively low electron affinity, 72.77 kJ/mol and reacts exothermically with protons as a powerful Lewis base.

$$H^- + H^+ \rightarrow H_2; \Delta H = -1676 \text{ kJ/mol}$$

The low electron affinity of hydrogen and the strength of the H–H bond (ΔH_{BE} = 436 kJ/mol) means that the hydride ion would also be a strong reducing agent

$$H_2 + 2e^- \rightleftharpoons 2H^-; E^{\circ} = -2.25 \text{ V}$$

Types of Hydrides

According to the general definition every element of the periodic table (except some noble gases) forms one or more hydrides. These substances have been classified into three main types according to the nature of their bonding:

- *Ionic hydrides*, which have significant ionic bonding character.

- *Covalent hydrides*, which include the hydrocarbons and many other compounds which covalently bond to hydrogen atoms.

- *Interstitial hydrides*, which may be described as having metallic bonding.

While these divisions have not been used universally, they are still useful to understand differences in hydrides.

Ionic Hydrides

Ionic or saline hydrides are composed of hydride bound to an electropositive metal, generally an alkali metal or alkaline earth metal. The divalent lanthanides such as europium and ytterbium form compounds similar to those of heavier alkali metal. In these materials the hydride is viewed as a pseudohalide. Saline hydrides are insoluble in conventional solvents, reflecting their non-molecular structures. Ionic hydrides are used as bases and, occasionally, as reducing reagents in organic synthesis.

$$C_6H_5C(O)CH_3 + KH \rightarrow C_6H_5C(O)CH_2K + H_2$$

Typical solvents for such reactions are ethers. Water and other protic solvents cannot serve as a medium for ionic hydrides because the hydride ion is a stronger base than hydroxide and most hydroxyl anions. Hydrogen gas is liberated in a typical acid-base reaction.

$$NaH + H_{2O} \rightarrow H_2 \text{ (g)} + NaOH \quad \Delta H = -83.6 \text{ kJ/mol}, \Delta G = -109.0 \text{ kJ/mol}$$

Often alkali metal hydrides react with metal halides. Lithium aluminium hydride (often abbreviated as LAH) arises from reactions of lithium hydride with aluminium chloride.

$$4 \text{ LiH} + AlCl_3 \rightarrow LiAlH_4 + 3 \text{ LiCl}$$

Covalent Hydrides

According to some definitions, covalent hydrides cover all other compounds containing hydrogen. Some definitions limit hydrides to hydrogen centres that formally react as hydrides, i.e. are nucleophilic, and hydrogen atoms bound to metal centers.These hydrides are formed by all the true non-metals (except zero group elements) and the elements like Al, Ga, Sn, Pb, Bi, Po, etc., which are normally metallic in nature, i.e., this class includes the hydrides of p-block elements. In these substances the hydride bond is formally a covalent bond much like the bond made by a proton in a weak acid. This category includes hydrides that exist as discrete molecules, polymers or oligomers, and hydrogen that has been chem-adsorbed to a surface. A particularly important segment of covalent hydrides are complex metal hydrides, powerful soluble hydrides commonly used in synthetic procedures.

Molecular hydrides often involve additional ligands such as, diisobutylaluminium hydride (DIBAL) consists of two aluminum centers bridged by hydride ligands. Hydrides that are soluble in common solvents are widely used in organic synthesis. Particularly common are sodium borohydride ($NaBH_4$) and lithium aluminium hydride and hindered reagents such as DIBAL.

Interstitial Hydrides or Metallic Hydrides

Interstitial hydrides most commonly exist within metals or alloys. They are traditionally termed 'compounds', even though they do not strictly conform to the definition of a compound; more closely resembling common alloys such as steel. In such hydrides, hydrogen can exist as either atomic, or diatomic entities. Mechanical or thermal processing, such as bending, striking, or annealing may cause the hydrogen to precipitate out of solution, by degassing. Their bonding is generally considered metallic. Such bulk transition metals form interstitial binary hydrides when exposed to hydrogen. These systems are usually non-stoichiometric, with variable amounts of hydrogen atoms in the lattice. In materials engineering, the phenomenon of hydrogen embrittlement results from the formation of interstitial hydrides. Hydrides of this type form according to either one of two main mechanisms. The first mechanism involves the adsorption of dihydrogen, succeeded by the cleaving of the H-H bond, the delocalisation of the hydrogen's electrons, and finally, the diffusion of the protons into the metal lattice. The other main mechanism involves the electrolytic reduction of ionised hydrogen on the surface of the metal lattice, also followed by the diffusion of the protons into the lattice. The second mechanism is responsible for the observed temporary volume expansion of certain electrodes used in electrolytic experiments.

Palladium absorbs up to 900 times its own volume of hydrogen at room temperatures, forming palladium hydride. This material has been discussed as a means to carry hydrogen for vehicular fuel cells. Interstitial hydrides show certain promise as a way for safe hydrogen storage. During last 25 years many interstitial hydrides were developed that readily absorb and discharge hydrogen at room temperature and atmospheric pressure. They are usually based on intermetallic compounds and solid-solution alloys. However, their application is still limited, as they are capable of storing only about 2 weight percent of hydrogen, insufficient for automotive applications.

Transition Metal Hydride Complexes

Transition metal hydrides include compounds that can be classified as *covalent hydrides*. Some are even classified as interstitial hydrides and other bridging hydrides. Classical transition metal hydride feature a single bond between the hydrogen centre and the transition metal. Some transition metal hydrides are acidic, e.g., $HCo(CO)_4$ and $H_2Fe(CO)_4$. The anions $[ReH_9]^{2-}$ and $[FeH_6]^{4-}$ are examples from the growing collection of known molecular homoleptic metal hydrides. As pseudohalides, hydride ligands are capable of bonding with positively polarized hydrogen centres. This interaction, called dihydrogen bond is similar to hydrogen bonding which exists between positively polarized protons and electronegative atoms with open lone pairs.

Deuterides

Hydrides containing deuterium are known as *deuterides*. Some deuterides, such as LiD, are important fusion fuels in thermonuclear weapons, and useful moderators in nuclear reactors.

Appendix on Nomenclature

Protide, *deuteride*, and *tritide* are used to describe ions or compounds, which contain enrichedhydrogen-1, deuterium or tritium, respectively.

In the classic meaning, hydride refers to any compounds hydrogen forms with other elements, ranging over groups 1–16 (the binary compounds of hydrogen). The following is a list of the nomenclature for the hydride derivatives of main group compounds according to this definition:

- alkali and alkaline earth metals: metal hydride

- boron: borane, BH_3

- aluminium: alumane, AlH_3

- gallium: gallane, GaH_3

- indium: indigane, InH_3

- thallium: thallane, TlH_3

- carbon: alkanes, alkenes, alkynes, and all hydrocarbons

- silicon: silane

- germanium: germane

- tin: stannane

- lead: plumbane

- nitrogen: ammonia ('azane' when substituted), hydrazine

- phosphorus: phosphine (note 'phosphane' is the IUPAC recommended name)

- arsenic: arsine (note 'arsane' is the IUPAC recommended name)

- antimony: stibine (note 'stibane' is the IUPAC recommended name)

- bismuth: bismuthine (note 'bismuthane' is the IUPAC recommended name)

- helium: helium hydride (only exists as an ion)

According to the convention above, the following are "hydrogen compounds" and not "hydrides":

- oxygen: water ('oxidane' when substituted; synonym: oxygen hydride), hydrogen peroxide

- sulfur: hydrogen sulfide ('sulfane' when substituted) synonym: sulfur hydride

- selenium: hydrogen selenide ('selane' when substituted)

- tellurium: hydrogen telluride ('tellane' when substituted)

- polonium: hydrogen polonide ('polane' when substituted)

- halogens: hydrogen halides

Examples:

- nickel hydride: used in NiMH batteries

- palladium hydride: electrodes in cold fusion experiments

- lithium aluminium hydride: a powerful reducing agent used in organic chemistry

- sodium borohydride: selective specialty reducing agent, hydrogen storage in fuel cells

- sodium hydride: a powerful base used in organic chemistry

- diborane: reducing agent, rocket fuel, semiconductor dopant, catalyst, used in organic synthesis; also borane, pentaborane and decaborane

- arsine: used for dopingsemiconductors

- stibine: used in semiconductor industry

- phosphine: used for fumigation

- silane: many industrial uses, e.g. manufacture of composite materials and water repellents

- ammonia: coolant, fuel, fertilizer, many other industrial uses

- hydrogen sulfide: component of natural gas, important source of sulfur

- Chemically, even water and hydrocarbons could be considered hydrides.

All metalloid hydrides are highly flammable. All solid non-metallic hydrides except ice are highly flammable. But,when Hydrogen combines with halogens, it produces acids rather than hydrides and they are not flammable.

Precedence Convention

According to IUPAC convention, by precedence (stylized electronegativity), hydrogen falls between group 15 and group 16 elements. Therefore, we have NH_3, 'nitrogen hydride' (ammonia), versus H_2O, 'hydrogen oxide' (water). This convention is sometimes broken for polonium, which on the grounds of polonium's metallicity is often referred to as 'polonium hydride' instead of the expected 'hydrogen polonide'.

Synthesis

Following reactions are employed for synthesizing metal hydrides.

i.   *Protonation reactions*

For this reaction to occur the metal center has to be basic and electron rich.

$$\left[Fe(CO)_4\right]^{2-} \xrightarrow{\ H^+\ } \left[HFe(CO)_4\right]^- \xrightarrow{\ H^+\ } H_2Fe(CO)_4$$

ii.   *From hydride donors*

Generally for this method, a main group hydride is reacted with metal halide.

$$WCl_6 + LiBEt_3H + PR_3 \rightarrow WH_6(PR_3)_3$$

iii.   *Using dihydrogen (H2) addition*

This method involves oxidative addition of H2 and thus requires metal centers that are capable of undergoing the oxidative addition step.

$$WMe_6 + PMe_2Ph \xrightarrow{H_2} WH_6(PMe_2Ph)_3$$

iv.   *From a ligand*

This method takes into account the β−elimination that occur in a variety of metal bound ligand moieties, thereby yielding a M−H bond.

$$RuCl_2(PPh_3)_3 + KOCHMe_2 + PPh_3 \rightarrow RuH_2(PPh_3)_4 + Me_2CO + KCl$$

Reactions of Metal Hydrides

Metal hydrides are reactive species kinetically and thus participate in a variety of transformations like the ones discussed below.

i.   *Deprotonation reactions*

The deprotonation reaction can be achieved by a hydride moiety resulting in the formation of H2 gas as shown below.

$$WH_6(PMe_3)_3 + NaH \rightarrow Na\left[WH_5(PMe_3)_3\right] + H_2$$

ii.   Hydride transfer and insertion

In this reaction a hydride transfer from a metal center to formaldehyde resulting in the formation of a metal bound methoxy moiety is observed as shown below.

$$Cp_2^+ZrH_2 + CH_2O \rightarrow Cp_2^+Zr(OMe)_2$$

iii.   Hydrogen atom transfer reaction

An example of hydrogen atom transfer reaction is given below.

$$\left[Co(CN)_5H\right]^{3-} + PhCH = CHCOOH \rightarrow \left[Co(CN)_5\right]^{3-} + PhCH - CH_2COOH$$

It is interesting to note that the nature of hydrogen atom in a M–H bond can vary from being protic in nature, when bound to electron deficient metal centers as in metal carbonyl compounds, to that of being hydridic in nature, when bound to more electropositive early transition metals. In the latter case, the hydride moieties tend to be basic and exhibit hydride transfer reactions with electrophiles like aldehydes or ketones. Furthermore, the protonation of these basic metal hydrides leads to the elimination of dihydrogen (H_2) gas along with the generation of a vacant coordination site at the metal center.

Bridging Hydrides

The metal hydrides usually show two modes of binding, namely terminal and bridging. In case of the bridging hydrides, the hydrogen atom can bridge between two or even more metal centers and thus, the bridging hydrides often display bent geometries.

σ–Complexes

σ–complexes are rare compounds, in which the σ bonding electrons of a X–H bond further participate in bonding with a metal center (X = H, Si, Sn, B, and P). The σ complexes thus exhibit an askewed binding to a metal center with the hydrogen atom, containing no lone pair, being more close to the metal center and thereby resulting in a side–on structure. Many times if the metal center is electron rich, then further back donation to the σ* orbital of the metal bound X–H moiety may occur resulting in a complete cleavage of the X–H bond.

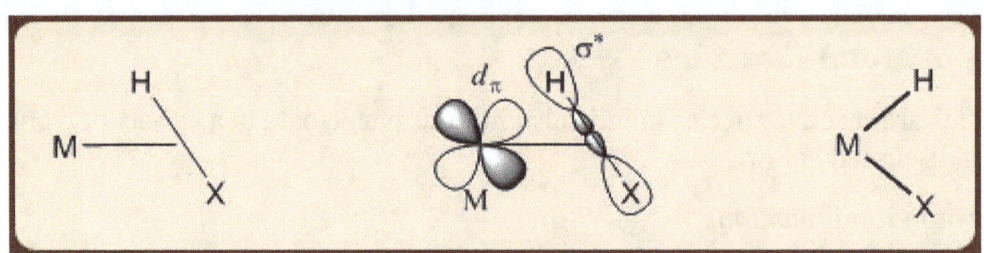

Metal Dihydrogen Complexes

The simplest variant of a σ−complex contains a dihydrogen ligand. The first dihydrogen complex was isolated by Kubas, after which many new ones were reported.

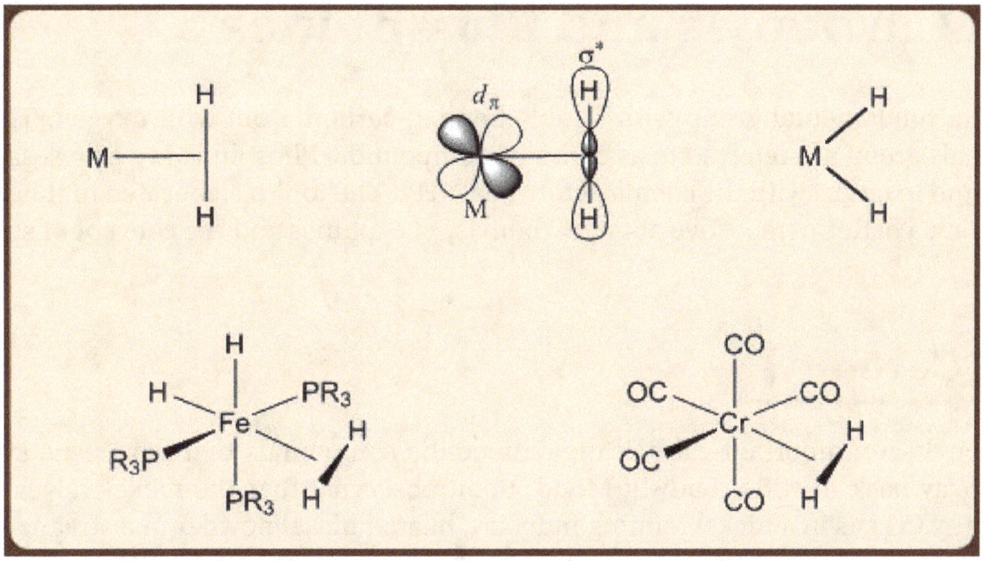

Quite expectedly, the dihydrogen moiety bound to a metal in a σ−complex is found to be more acidic (pKa = 0 – 20) when compared to the free dihydrogen molecule (pKa = 35). It is interesting to note that the pKa change associated with the binding of dihydrogen to a metal in a σ−complex relative to that of the free H_2 molecule is significantly larger than the change associated with binding of H_2O to metal. Owing to this inherent acidity, the deprotonation of the metal bound dihydrogen moiety by a base can thus be appropriately employed for heterolytic activation of the dihydrogen moiety as illustrated below.

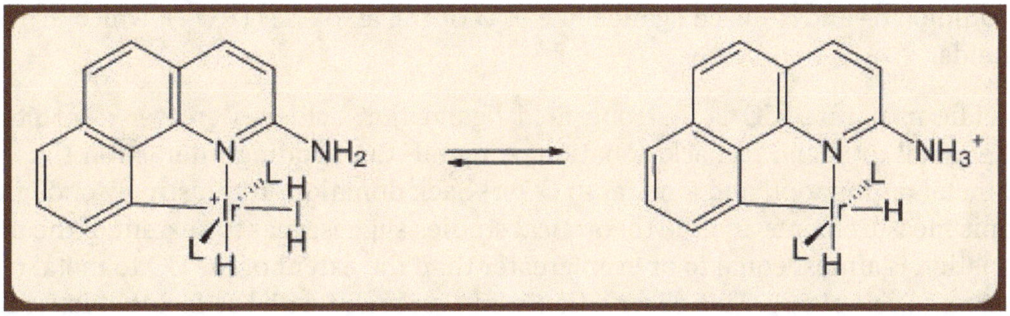

The dihydrogen complexes of metals are often referred to as nonclassical hydrides. The electron rich π basic metals are anticipated to split the metal bound dihydrogen moieties resulting in classical dihydride complexes. Along the same line of thinking, the electron deficient and less π basic metal would tend to stabilize a dihydrogen complex. The dihydrogen complexes can also be characterized by the X ray diffraction as well as neutron diffraction methods. In IR spectrum, the metal bound H−H stretch appear in the range (2300 – 2900) cm⁻¹ while in the ¹H NMR spectrum the same appear between 0 to −10 ppm as a broad peak. The dihydrogen complexes are often characterized by isotopic labeling studies of metal bound H−D moiety that shows a coupling constant of 20 – 34 Hz as supposed to 43 Hz observed in case of the free H−D molecule.

6

Metal Carbonyls and Phosphines

Carbonyl is a fundamental group with double-bonded carbon atom with oxygen. The elements containing this group are referred to as carbonyl compounds. Phosphine is a colorless, odourless, flammable and toxic gas with the chemical formula NH3. The topics elaborated in this chapter will help in gaining a better perspective about carbonyls, phosphines and the concept of substitution.

Metal Carbonyl

Metal carbonyls are important class of organometallic compounds that have been studied for a long time. Way back in 1884, Ludwig Mond, upon observing that the nickel valves were being eating away by CO gas in a nickel refining industry, heated nickel powder in a stream of CO gas to synthesize the first known metal carbonyl compound in the form $Ni(CO)_4$. The famous Mond refining process was thus born, grounded on the premise that the volatile $Ni(CO)_4$ compound can be decomposed to pure metal at elevated temperature. Mond subsequently founded the Mond Nickel Company Limited for purifying nickel from its ore using this method.

The carbonyl ligand (CO) distinguishes itself from other ligands in many respects. For example, unlike the alkyl ligands, the carbonyl (CO) ligand is unsaturated thus allowing not only the ligand to σ−donate but also to accept electrons in its π^* orbital from d_π metal orbitals and thereby making the CO ligand π−acidic. The other difference lies in the fact that CO is a soft ligand compared to the other common σ−and π−basic ligands like H_2O or the alkoxides (RO−), which are considered as hard ligands.

Being π−acidic in nature, CO is a strong field ligand that achieves greater d−orbital splitting through the metal to ligand π−back donation. A metal–CO bonding interaction thus comprises of a CO to metal σ−donation and a metal to CO π−back donation. Interestingly enough, both the spectroscopic measurements and the theoretical studies suggest that the extent of the metal to CO π−back donation is almost equal to or even greater than the extent of the CO to metal σ−donation in metal carbonyl complexes. This observation is in agreement with the fact that low valent−transition metal centers tend to form metal carbonyl complexes.

Metal Carbonyls

Metal carbonyls are coordination complexes of transition metals with carbon monoxideligands. Metal carbonyls are useful in organic synthesis and as catalysts or catalyst precursors in homogeneous catalysis, such as hydroformylation and Reppe chemistry. In the Mond process, nickel carbonyl is used to produce pure nickel. In organometallic chemistry, metal carbonyls serve as precursors for the preparation of other organometalic complexes.

Metal carbonyls are toxic by skin contact, inhalation or ingestion, in part because of their ability to carbonylate hemoglobin to give carboxyhemoglobin, which prevents the binding of O_2.

Nomenclature and Terminology

The nomenclature of the metal carbonyls depends on the charge of the complex, the number and type of central atoms, and the number and type of ligands and their binding modes. They occur as neutral complexes, as positively charged metal carbonyl cations or as negatively charged metal carbonylates. The carbon monoxide ligand may be bound terminally to a single metal atom or bridging to two or more metal atoms. These complexes may be homoleptic, that is containing only CO ligands, such as nickel carbonyl ($Ni(CO)_4$), but more commonly metal carbonyls are heteroleptic and contain a mixture of ligands.

Mononuclear metal carbonyls contain only one metal atom as the central atom. Except vanadium hexacarbonyl only metals with even order number such as chromium, iron, nickel and their homologs build neutral mononuclear complexes. Polynuclear metal carbonyls are formed from metals with odd order numbers and contain a metal-metal bond. Complexes with different metals, but only one type of ligand will be referred to as isoleptic.

The number of carbon monoxide ligands in a metal carbonyl complex is described by a Greek numeral, followed by the word *carbonyl*. Carbon monoxide has different binding modes in metal carbonyls. They differ in the hapticity and the bridging mode. The hapticity describes the number of carbon monoxide ligands, which are directly bonded to the central atom. The denomination shall be made by the letter η^n, which is prefixed to the name of the complex. The superscript n indicates the number of bounded atoms. In monohapto coordination, such as in terminally bonded carbon monoxide, the hapticity is 1 and it is usually not separately designated. If carbon monoxide is bound via the carbon atom and via the oxygen to the metal, it will be referred to as dihapto coordinated η^2.

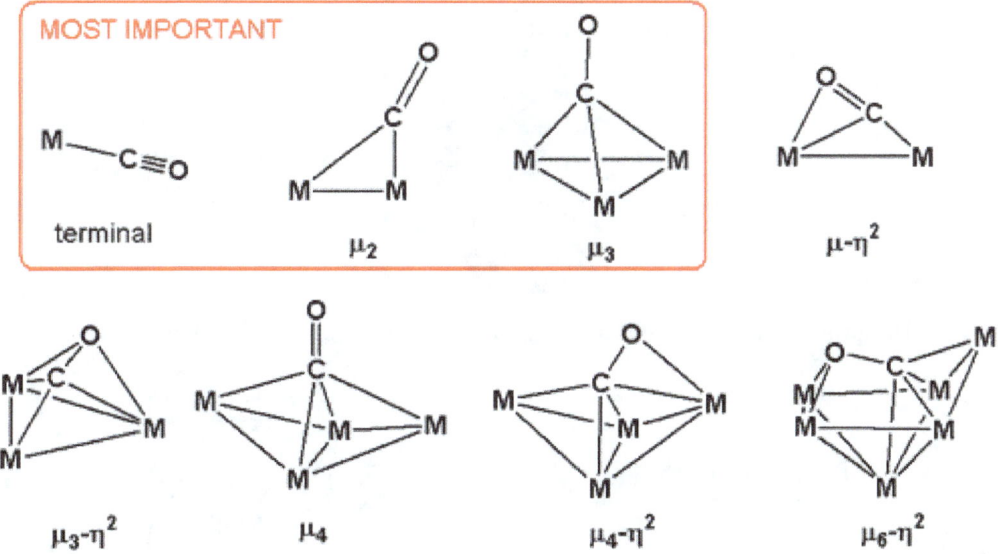

The carbonyl ligand engages in a range of bonding modes in metal carbonyl dimers and clusters. In the most common bridging mode, the CO ligand bridges a pair of metals. This bonding mode is observed in the commonly available metal carbonyls: $Co_2(CO)_8$, $Fe_2(CO)_9$, $Fe_3(CO)_{12}$, and $Co_4(CO)_{12}$. In certain higher nuclearity clusters, CO bridges between three or even four metals. These ligands are denoted μ_3-CO and μ_4-CO. Less common are bonding modes in which both C and O bond to the metal, e.g. μ_3-η^2.

Structure and Bonding

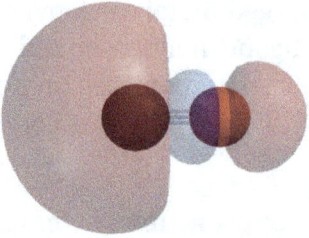

The highest occupied molecular orbital (HOMO) of CO is a σ MO

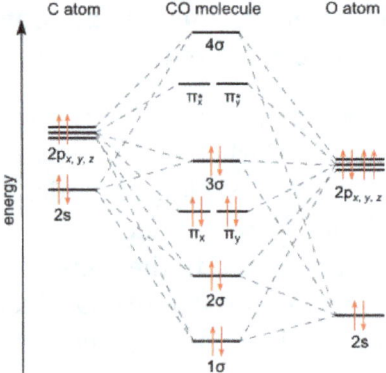

Energy level scheme of the σ and π orbitals of carbon monoxide

The lowest unoccupied molecular orbital (LUMO) of CO is a π* antibondingMO

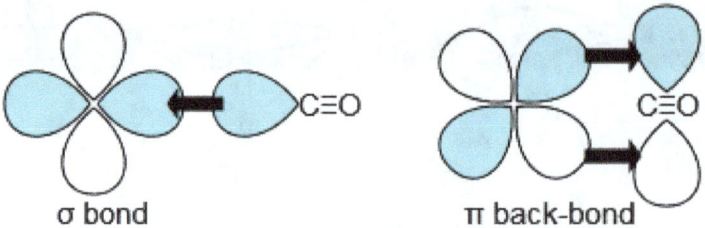

Diagram showing synergic *π back-bonding* in transition metal carbonyls

Carbon monoxide bonds to transition metals using "synergistic π* back-bonding." The bonding has three components, giving rise to a partial triple bond. A sigma bond arises from overlap of the nonbonding (or weakly anti-bonding) sp-hybridized electron pair on carbon with a blend of d-, s-, and p-orbitals on the metal. A pair of π bonds arises from overlap of filled d-orbitals on the metal with a pair of π-antibonding orbitals projecting from the carbon atom of the CO. The latter kind of binding requires that the metal have d-electrons, and that the metal is in a relatively low oxidation state (<+2) which makes the back donation process favorable. As electrons from the metal fill the π-antibonding orbital of CO, they weaken the carbon-oxygen bond compared with free carbon monoxide, while the metal-carbon bond is strengthened. Because of the multiple bond character of the M-CO linkage, the distance between the metal and carbon atom is relatively short, often < 1.8 Å, about 0.2 Å shorter than a metal-alkyl bond. Several canonical forms can be drawn to describe the *approximate* metal carbonyl bonding modes.

$$M^- - C \equiv O^+ \leftrightarrow M = C = O \leftrightarrow M^+ \equiv C - O^-$$

Resonance structures of a metal carbonyl, from left to right the contributions of the right-hand-side canonical forms increase as the back bonding power of M to CO increases.

Infrared spectroscopy is a sensitive probe for the presence of bridging carbonyl ligands. For compounds with doubly bridging CO ligands, denoted μ_2-CO or often just μ-CO, v_{CO}, v_{CO} is usually shifted by 100–200 cm^{-1} to lower energy compared to the signatures of terminal CO, i.e. in the region 1800 cm^{-1}. Bands for face capping (μ_3) CO ligands appear at even lower energies. Typical values for rhodium cluster carbonyls are: In addition to symmetrical bridging modes, CO can be found bridge unsymmetrically or through donation from a metal d orbital to the π* orbital of CO. The increased π-bonding due to back-donation from multiple metal centers results in further weakening of the C-O bond.

Physical Characteristics

Most mononuclear carbonyl complexes are colorless or pale yellow volatile liquids or solids that are flammable and toxic. Vanadium hexacarbonyl, a uniquely stable 17-electron metal carbonyl, is a blue-black solid. Di- and polymetallic carbonyls tend to be more deeply colored. Triiron dodecacarbonyl ($Fe_3(CO)_{12}$) forms deep green crystals. The crystalline metal carbonyls often are sublimable in vacuum, although this process is often accompanied by degradation. Metal carbonyls are soluble in nonpolar and polar organic solvents such as benzene, diethyl ether, acetone, glacial acetic acid and carbon tetrachloride. Some salts of cationic and anionic metal carbonyls are soluble in water or lower alcohols.

Analytical Characterization

Isomers of dicobalt octacarbonyl

Apart from X-ray crystallography, important analytical techniques for the characterization of metal carbonyls are infrared spectroscopy and ^{13}C NMR spectroscopy. These two techniques provide structural information on two very different time scales. Infrared active vibrational modes, such as CO-stretching vibrations are often fast compared to intramolecular processes, whereas NMR transitions occur at lower frequencies and thus sample structures on a time scale that, it turns out, is comparable to the rate of intramolecular ligand exchange processes. NMR data provide information on "time-averaged structures," whereas IR is an instant "snapshot." Illustrative of the differing time scales, investigation of dicobalt octacarbonyl ($Co_2(CO)_8$) by means of infrared spectroscopy provides 13 v_{CO} bands, far more than expected for a single compound. This complexity reflects the presence of isomers with and without bridging CO-ligands. The ^{13}C-NMR spectrum of the same substance exhibits only a single signal at a chemical shift of 204 ppm. This simplicity indicates that the isomers quickly (on the NMR timescale) interconvert.

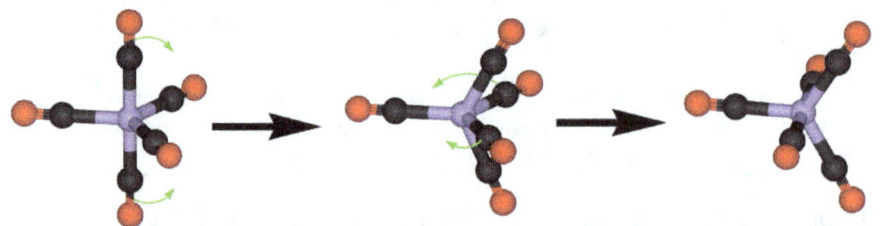

The Berry pseudorotation mechanism for iron pentacarbonyl

Iron pentacarbonyl exhibits only a single ^{13}C-NMR signal owing to rapid exchange of the axial and equatorial CO ligands by Berry pseudorotation.

Infrared Spectra

The most important technique for characterizing metal carbonyls is infra-red spectroscopy. The C-O vibration, typically denoted v_{CO}, occurs at 2143 cm^{-1} for CO gas. The energies of the v_{CO} band for the metal carbonyls correlates with the strength of the carbon-oxygen bond, and inversely correlated with the strength of the π-backbonding between the metal and the carbon. The π basicity of the metal center depends on a lot of factors; in the isoelectronic series (Ti to Fe) at the bottom of this chapter, the hexacarbonyls show decreasing π-backbonding as one increases (makes more positive) the charge on the metal. π-Basic ligands increase π-electron density at the metal, and improved backbonding reduces v_{CO}. The Tolman electronic parameter uses the $Ni(CO)_3$ fragment to order ligands by their π-donating abilities.

The number of vibrational modes of a metal carbonyl complex can be determined by group theory. Only vibrational modes that transform as the electric dipole operator will have non-zero direct products and are observed. The number of observable IR transitions (but not the energies) can thus be predicted. For example, the CO ligands of octahedral complexes, e.g. $Cr(CO)_6$, transform as a_{1g}, e_g, and t_{1u}, but only the t_{1u} mode (anti-symmetric stretch of the apical carbonyl ligands) is IR-allowed. Thus, only a single v_{CO} band is observed in the IR spectra of the octahedral metal hexacarbonyls. Spectra for complexes of lower symmetry are more complex. For example, the IR spectrum of $Fe_2(CO)_9$ displays CO bands at 2082, 2019, 1829 cm^{-1}. Exhaustive tabulations are available.

Coordination number

	4	5	6
3 Carbonyl ligands			
IR-Peaks	2	1	2
IR-Peaks		3	3
IR-Peaks		3	
4 Carbonyl ligands			
IR-Peaks	1	4	1
IR-Peaks		3	4
5 Carbonyl ligands			
IR-Peaks		2	3
6 Carbonyl ligands			
IR-Peaks			1

The number of IR-active vibrational modes of several prototypical metal carbonyl complexes.

Compound	v_{CO} (cm^{-1})	^{13}C NMR shift
CO	2143	181
$Ti(CO)_6^{-2}$	1748	
$V(CO)_6^{-1}$	1859	
$Cr(CO)_6$	2000	212
$Mn(CO)_5^+$	2100	
$Fe(CO)_4^{2+}$	2204	
$Fe(CO)_5$	2022, 2000	209

carbonyl	v_{CO}, μ_1 (cm^{-1})	v_{CO}, μ_2 (cm^{-1})	v_{CO}, μ_3 (cm^{-1})
$Rh_2(CO)_8$	2060, 2084	1846, 1862	
$Rh_4(CO)_{12}$	2044, 2070, 2074	1886	
$Rh_6(CO)_{16}$	2045, 2075		1819

Nuclear Magnetic Resonance Spectroscopy

The traditional method for the study of metal carbonyls is the ^{13}C NMR spectroscopy. To improve the sensitivity of this technique, complexes are often enriched ^{13}CO. Typical range for the chemical shift for terminally bound ligands is 150 to 220 ppm. Bridging ligands absorb between 230 and 280 ppm. The ^{13}C signals shift toward higher fields with an increasing atomic number of the central metal.

The nuclear magnetic resonance spectroscopy can be used for experimental determination of the complex dynamics. The activation energy of ligand exchanges processes can be determined by the temperature dependence of the line broadening.

Mass Spectrometry

Mass spectrometry provides information about the structure and composition of the complexes. Spectra for metal polycarbonyls are often easily interpretable, because the dominant fragmentation process is the loss of carbonyl ligands (m/z = 28).

$$M(CO)_n^+ \rightarrow M(CO)_{n-1}^+ + CO$$

Electron impact ionization is the most common technique for characterizing the neutral metal carbonyls. Neutral metal carbonyls can be converted to charged species by derivatization, which enables the use of electrospray ionization, instrumentation for which is often widely available. For example, treatment of a metal carbonyl with alkoxide generates an anionic metallaformate that is amenable to analysis by ESI-MS:

$$L_nM(CO) + RO^- \rightarrow [L_nM\text{-}C(=O)OR]^-$$

Some metal carbonyls react with azide to give isocyanato complexes with release of nitrogen. By adjusting the cone voltage and/or temperature, the degree of fragmentation can be controlled. The molar mass of the parent complex can be determined, as well as information about structural rearrangements involving loss of carbonyl ligands under ESI-MS conditions.

Occurrence in Nature

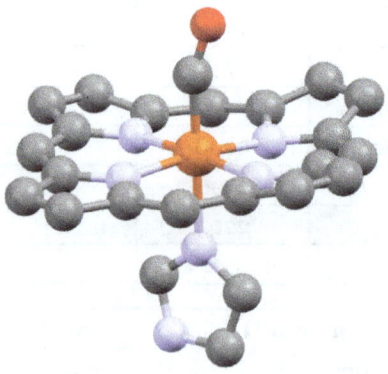

A heme unit of human carboxyhemoglobin, showing the carbonyl ligand at the apical position, *trans* to the histidine residue.

In the investigation of the infrared spectrum of the Galactic Center monoxide vibrations of iron carbonyls in interstellar dust clouds were detected. Iron carbonyl clusters were also observed in Jiange H5 chondrites identified by infrared spectroscopy. Four infrared stretching frequencies were found for the terminal and bridging carbon monoxide ligands.

In the oxygen-rich atmosphere of earth metal carbonyls are subject to oxidation to the metal oxides. It is discussed whether in the reducing hydrothermal environments of the pre-biotic prehistory such complexes were formed and could have been available as catalysts for the synthesis of critical biochemical compounds such as pyruvic acid. Traces of the carbonyls of iron, nickel, and tungsten were found in the gaseous emanations from the sewage sludge of municipal treatment plants.

The hydrogenase enzymes contain CO bound to iron. Apparently the CO stabilizes low oxidation states, which facilitates the binding of hydrogen. The enzymes carbon monoxide dehydrogenase and acetyl coA synthase also are involved in bio-processing of CO. Carbon monoxide containing complexes are invoked for the toxicity of CO and signaling.

Synthesis

The synthesis of metal carbonyls is subject of intense organometallic research. Since the work of Mond and then Hieber, many procedures have been developed for the preparation of mononuclear metal carbonyls as well as homo-and hetero-metallic carbonyl clusters.

Direct Reaction of Metal with Carbon Monoxide

Nickel tetracarbonyl and iron pentacarbonyl can be prepared according to the following equations by reaction of finely divided metal with carbon monoxide:

$$Ni + 4\,CO \rightarrow Ni(CO)_4 \text{ (1 bar, 55 °C)}$$

$$Fe + 5\,CO \rightarrow Fe(CO)_5 \text{ (100 bar, 175 °C)}$$

Nickel carbonyl is formed with carbon monoxide already at 80 °C and atmospheric pressure, finely divided iron reacts at temperatures between 150 and 200 °C and a carbon monoxide pressure of 50 to 200 bar. Other metal carbonyls are prepared by less direct methods.

Reduction of Metal Salts and Oxides

Some metal carbonyls are prepared by the reduction of metal halides in the presence of high pressure of carbon monoxide. A variety of reducing agents are employed, including copper, aluminum, hydrogen, as well as metal alkyls such as triethylaluminum. Illustrative is the formation of chromium hexacarbonyl from anhydrous chromium(III) chloride in benzene with aluminum as a reducing agent, and aluminum chloride as the catalyst:

$$CrCl_3 + Al + 6\,CO \rightarrow Cr(CO)_6 + AlCl_3$$

The use of metal alkyls, e.g. triethylaluminium and diethylzinc as the reducing agent leads to the oxidative coupling of the alkyl radical to the dimer:

$$WCl_6 + 6\,CO + 2\,Al(C_2H_5)_3 \rightarrow W(CO)_6 + 2\,AlCl_3 + 3\,C_4H_{10}$$

Tungsten, molybdenum, manganese, and rhodium salts may be reduced with lithium aluminum hydride. Vanadium hexacarbonyl is prepared with sodium as a reducing agent in chelating solvents such as diglyme.

$$VCl_3 + 4\,Na + 6\,CO\ 2\ diglyme \rightarrow Na(diglyme)_2[V(CO)_6] + 3\,NaCl$$

$$[V(CO)_6]^- + H^+ \rightarrow H[V(CO)_6] \rightarrow 1/2\,H_2 + V(CO)_6$$

In aqueous phase nickel or cobalt salts can be reduced, for example, by sodium dithionite. In the presence of carbon monoxide, cobalt salts are quantitatively converted to the tetracarbonylcobalt(-1) anion:

$$Co^{2+} + 1.5\,S_2O_4^{2-} + 6\,OH^- + 4\,CO \rightarrow Co(CO)_4^- + 3\,SO_3^{2-} + 3\,H_2O$$

Some metal carbonyls are prepared using CO as the reducing agent. In this way, Hieber and Fuchs first prepared dirhenium decacarbonyl from the oxide:

$$Re_2O_7 + 17\,CO \rightarrow Re_2(CO)_{10} + 7\,CO_2$$

If metal oxides are used carbon dioxide is formed as a reaction product. In the reduction of metal chlorides with carbon monoxide phosgene is formed, as in the preparation of osmium carbonyl chloride from the chloride salts. Carbon monoxide is also suitable for the reduction of sulfides, where carbonyl sulfide is the byproduct.

Photolysis and Thermolysis

Photolysis or thermolysis of mononuclear carbonyls generates bi- and multimetallic carbonyls such as diiron nonacarbonyl ($Fe_2(CO)_9$). On further heating, the products decompose eventually into the metal and carbon monoxide.

$$2\,Fe(CO)_5 \rightarrow Fe_2(CO)_9 + CO$$

The thermal decomposition of triosmium dodecacarbonyl ($Os_3(CO)_{12}$) provides higher-nuclear osmium carbonyl clusters such as $Os_4(CO)_{13}$, $Os_6(CO)_{18}$ up to $Os_8(CO)_{23}$.

Mixed ligand carbonyls of ruthenium, osmium, rhodium, and iridium are often generated by abstraction of CO from solvents such as dimethylformamide (DMF) and 2-methoxyethanol. Typical is the synthesis of $IrCl(CO)(PPh_3)_2$ from the reaction of iridium(III) chloride and triphenylphosphine in boiling DMF solution.

Salt Metathesis

Salt metathesis reaction of for example $KCo(CO)_4$ with $[Ru(CO)_3Cl_2]_2$ leads selectively to mixed-metal carbonyls such as $RuCo_2(CO)_{11}$.

$$4\,KCo(CO)_4 + [Ru(CO)_3Cl_2]_2 \rightarrow 2\,RuCo_2(CO)_{11} + 4\,KCl + 11\,CO$$

Metal Carbonyl Cations and Carbonylates

The synthesis of ionic carbonyl complexes is possible by oxidation or reduction of the neutral complexes. Anionic metal carbonylates can be obtained for example by reduction of dinuclear complexes with sodium. A familiar example is the sodium salt of iron tetracarbonylate ($Na_2Fe(CO)_4$, *Collman's reagent*), which is used in organic synthesis.

The cationic hexacarbonyl salts of manganese, technetium and rhenium can be prepared from the carbonyl halides under carbon monoxide pressure by reaction with a Lewis acid.

$$Mn(CO)_5Cl + AlCl_3 + CO \rightarrow Mn(CO)_6^+AlCl_4^-$$

The use of strong acids succeeded in preparing gold carbonyl cations such as $[Au(CO)_2]^+$, which is used as a catalyst for the carbonylation of olefins. The cationic platinum carbonyl complex $[Pt(CO)_4]^+$ can be prepared by working in so-called super acids such as antimony pentafluoride.

Reactions

Metal carbonyls are important precursors for the synthesis of other organometalic complexes. The main reactions are the substitution of carbon monoxide by other ligands, the oxidation or reduction reactions of the metal center and reactions of carbon monoxide ligand.

CO Substitution

The substitution of CO ligands can be induced thermally or photochemically by donor ligands. The range of ligands is large, and includes phosphines, cyanide (CN^-), nitrogen donors, and even ethers, especially chelating ones. Olefins, especially diolefins, are effective ligands that afford synthetically useful derivatives. Substitution of 18-electron complexes generally follows a dissociative mechanism, involving 16-electron intermediates.

Substitution proceeds via a dissociative mechanism:

$$M(CO)_n \rightarrow M(CO)_{n-1} + CO$$

$$M(CO)_{n-1} + L \rightarrow M(CO)_{n-1}L$$

The dissociation energy is 105 kJ mol^{-1} for nickel carbonyl and 155 kJ mol^{-1} for chromium hexacarbonyl.

Substitution in 17-electron complexes, which are rare, proceeds via associative mechanisms with a 19-electron intermediates.

$$M(CO)_n + L \rightarrow M(CO)_nL$$

$$M(CO)_nL \rightarrow M(CO)_{n-1}L + CO$$

The rate of substitution in 18-electron complexes is sometimes catalysed by catalytic amounts of oxidants, via electron-transfer.

Reduction

Metal carbonyls react with reducing agents such as metallic sodium or sodium amalgam to give carbonylmetalate (or carbonylate) anions:

$$Mn_2(CO)_{10} + 2\,Na \rightarrow 2\,Na[Mn(CO)_5]$$

For iron pentacarbonyl, one obtains the tetracarbonylferrate with loss of CO:

$$Fe(CO)_5 + 2\,Na \rightarrow Na_2[Fe(CO)_4] + CO$$

Mercury can insert into the metal-metal bonds of some polynuclear metal carbonyls:

$$Co_2(CO)_8 + Hg \rightarrow (CO)_4Co\text{-}Hg\text{-}Co(CO)_4$$

Nucleophilic Attack at CO

The CO ligand is often susceptible to attack by nucleophiles. For example, trimethylamine oxide and bistrimethylsilylamide convert CO ligands to CO_2 and CN^-, respectively. In the "Hieber base reaction", hydroxide ion attacks the CO ligand to give a metallacarboxylic acid, followed by the release of carbon dioxide and the formation of metal hydrides or carbonylmetalates. A well-known example of this nucleophilic addition reaction is the conversion of iron pentacarbonyl to hydridorion tetracarbonyl anion:

$$Fe(CO)_5 + NaOH \rightarrow Na[Fe(CO)_4CO_2H]$$

$$Na[Fe(CO)_4COOH] + NaOH \rightarrow Na[HFe(CO)_4] + NaHCO_3$$

Protonation of the hydrido anion gives the neutral iron tetracarbonyl hydride:

$$Na[HFe(CO)_4] + H^+ \rightarrow H_2Fe(CO)_4 + Na^+$$

Organolithium reagents add with metal carbonyls to acylmetal carbonyl anions. O-alkylation of these anions, e.g. with Meerwein salts, affords Fischer carbenes.

With Electrophiles

Despite being in low formal oxidation states, metal carbonyls are relatively unreactive toward many electrophiles. For example, they resist attack by alkylating agents, mild acids, mild oxidizing agents. Most metal carbonyls do undergo halogenation. Iron pentacarbonyl, for example, forms ferrous carbonyl halides:

$$Fe(CO)_5 + X_2 \rightarrow Fe(CO)_4X_2 + CO$$

Metal-metal bonds are cleaved by halogens. Depending on the electron-counting scheme used, this can be regarded as oxidation of the metal atom:

$$Mn_2(CO)_{10} + Cl_2 \rightarrow 2\,Mn(CO)_5Cl$$

Compounds

Most metal carbonyl complexes contain a mixture of ligands. Examples include the historically important $IrCl(CO)(P(C_6H_5)_3)_2$ and the anti-knock agent $(CH_3C_5H_4)Mn(CO)_3$. The parent compounds for many of these mixed ligand complexes are the binary carbonyls, i.e. species of the formula $[M_x(CO)_n]^z$, many of which are commercially available. The formula of many metal carbonyls can be inferred from the 18 electron rule.

Charge-neutral Binary Metal Carbonyls

- Group 4 elements with 4 valence electrons are rare, but substituted derivatives of $Ti(CO)_7$ are known.

- Group 5 elements with 5 valence electrons, again are subject to steric effects that prevent the formation of M-M bonded species such as $V_2(CO)_{12}$, which is unknown. The 17 VE $V(CO)_6$ is however well known.

- Group 6 elements with 6 valence electrons form metal carbonyls $Cr(CO)_6$, $Mo(CO)_6$, and $W(CO)_6$ (6 + 6x2 = 18 electrons). Group 6 elements (as well as group 7) are well also well known for exhibiting the cis effect (the labilization of CO in the cis position) in organometallic synthesis.

- Group 7 elements with 7 valence electrons form metal carbonyl dimers $Mn_2(CO)_{10}$, $Tc_2(CO)_{10}$, and $Re_2(CO)_{10}$ (7 + 1 + 5x2 = 18 electrons).

- Group 8 elements with 8 valence electrons form metal carbonyls $Fe(CO)_5$, $Ru(CO)_5$ and $Os(CO)_5$ (8 + 5x2 = 18 electrons). The heavier two members are unstable, tending to decarbonylate to give $Ru_3(CO)_{12}$, and $Os_3(CO)_{12}$. The two other principal iron carbonyls are $Fe_3(CO)_{12}$ and $Fe_2(CO)_9$.

- Group 9 elements with 9 valence electrons and are expected to form metal carbonyl dimers $M_2(CO)_8$. In fact the cobalt derivative of this octacarbonyl is the only stable member, but all three tetramers are well known: $Co_4(CO)_{12}$, $Rh_4(CO)_{12}$, $Rh_6(CO)_{16}$, and $Ir_4(CO)_{12}$ (9 + 3 + 3x2 = 18 electrons). $Co_2(CO)_8$ unlike the majority of the other 18 VE transition metal carbonyls is sensitive to oxygen.

- Group 10 elements with 10 valence electrons form metal carbonyls $Ni(CO)_4$ (10 + 4x2 = 18 electrons). Curiously $Pd(CO)_4$ and $Pt(CO)_4$ are not stable.

Anionic Binary Metal Carbonyls

- Group 4 elements as dianions resemble neutral group 6 derivatives: $[Ti(CO)_6]^{2-}$.

- Group 5 elements as monoanions resemble again neutral group 6 derivatives: $[V(CO)_6]^-$.

- Group 7 elements as monoanions resemble neutral group 8 derivatives: $[M(CO)_5]^-$ (M = Mn, Tc, Re).

- Group 8 elements as dianaions resemble neutral group 10 derivatives: $[M(CO)_4]^{2-}$ (M = Fe, Ru, Os). Condensed derivatives are also known.

- Group 9 elements as monoanions resemble neutral group 10 metal carbonyl. $[Co(CO)_4]^-$ is the best studied member.

Large anionic clusters of Ni, Pd, and Pt are also well known.

Cationic Binary Metal Carbonyls

- Group 7 elements as monocations resemble neutral group 6 derivative $[M(CO)_6]^+$ (M = Mn, Tc, Re).

- Group 8 elements as dications also resemble neutral group 6 derivatives $[M(CO)_6]^{2+}$ (M = Fe, Ru, Os).

Metal Carbonyl Hydrides

Metal Carbonyl hydride	pK_a
$HCo(CO)_4$	"strong"
$HCo(CO)_3(P(OPh)_3)$	5.0
$HCo(CO)_3(PPh_3)$	7.0
$HMn(CO)_5$	7.1
$H_2Fe(CO)_4$	4.4, 14

Metal carbonyls are relatively distinctive in forming complexes with negative oxidation states. Examples include the anions discussed above. These anions can be protonated to give the corresponding metal carbonyl hydrides. The neutral metal carbonyl hydrides are often volatile and can be quite acidic.

Applications

Spheres of nickel manufactured by the Mond process

Metallurgical uses

Metal carbonyls are used in several industrial processes. Perhaps the earliest application was the extraction and purification of nickel via nickel tetracarbonyl by the Mond process.

By a similar process carbonyl iron, a highly pure metal powder, is prepared by thermal decomposition of iron pentacarbonyl. Carbonyl iron is used inter alia for the preparation of inductors, pigments, as dietary supplements, in the production of radar-absorbing materials in the stealth technology, and in Thermal spraying.

Catalysis

Metal carbonyls are used in a number of industrially important carbonylation reactions. In the oxo process, an olefin, dihydrogen, and carbon monoxide react together with a catalyst (e.g. dicobalt octacarbonyl) to give aldehydes. Illustrative is the production of butyraldehyde:

$$H_2 + CO + CH_3CH{=}CH_2 \rightarrow CH_3CH_2CH_2CHO$$

Butyraldehyde is converted on an industrial scale to 2-Ethylhexanol, a precursor to PVC plasticizers, by aldol condensation, followed by hydrogenation of the resulting hydroxyaldehyde. The "oxo aldehydes" resulting from hydroformylation are used for large-scale synthesis of fatty alcohols, which are precursors to detergents. The hydroformylation is a reaction with high atom economy, especially if the reaction proceeds with high regioselectivity.

Another important reaction catalyzed by metal carbonyls is the hydrocarboxylation. The example below is for the synthesis of acrylic acid and acrylic acid esters:

Also the cyclization of acetylene to cyclooctatetraene uses metal carbonyl catalysts:

$$4 \ \ HC\equiv CH \ \ \xrightarrow{\text{cat}}$$

In the Monsanto and Cativa processes, acetic acid is produced from methanol, carbon monoxide, and water using hydrogen iodide as well as rhodium and iridium carbonyl catalysts, respectively. Related carbonylation reactions afford acetic anhydride.

CO-releasing Molecules (CO-RMs)

Carbon monoxide-releasing molecules are metal carbonyl complexes that are being developed as potential drugs to release CO. At low concentrations, CO functions as a vasodilatory and an anti-inflammatory agent. CO-RMs have been conceived as a pharmacological strategic approach to carry and deliver controlled amounts of CO to tissues and organs.

Related Compounds

Many ligands are known to form homoleptic and mixed ligand complexes that are analogous to the metal carbonyls.

Nitrosyl Complexes

Metal nitrosyls, compounds featuring NO ligands, are numerous. In contrast to metal carbonyls, however, homoleptic metal nitrosyls are rare. NO is a stronger pi-acceptor than CO. Well known nitrosyl carbonyls include $CoNO(CO)_3$ and $Fe(NO)_2(CO)_2$, which are analogues of $Ni(CO)_4$.

Thiocarbonyls Complexes

Complexes containing CS are known but are uncommon. The rarity of such complexes is attributable in part to the fact that the obvious source material, carbon monosulfide, is unstable. Thus, the synthesis of thiocarbonyl complexes requires more elaborate routes, such as the reaction of disodium tetracarbonylferrate with thiophosgene:

$$Na_2Fe(CO)_4 + CSCl_2 \rightarrow Fe(CO)_4CS + 2 \ NaCl$$

Complexes of CSe and CTe are very rare.

Phosphine Complexes

All metal carbonyls undergo substitution by organophosphorus ligands. For example, the series $Fe(CO)_{5-x}(PR_3)_x$ is well known for various phosphine ligands for x = 1, 2, and 3. PF_3 behaves similarly but is remarkable because it readily forms homoleptic analogues of the binary metal carbonyls. For example, the volatile, stable complexes $Fe(PF_3)_5$ and $Co_2(PF_3)_8$ represent CO-free analogues of $Fe(CO)_5$ and $Co_2(CO)_8$ (unbridged isomer).

Isocyanide Complexes

Isocyanides also form extensive families of complexes that are related to the metal carbonyls. Typical isocyanide ligands are methyl and t-butyl isocyanides (Me_3CNC). A special case is CF_3NC, an unstable molecule that forms stable complexes whose behavior closely parallels that of the metal carbonyls.

Toxicology

The toxicity of metal carbonyls is due to toxicity of carbon monoxide, the metal, and because of the volatility and instability of the complexes. Exposure occurs by inhalation, or for liquid metal carbonyls by ingestion or due to the good fat solubility by skin resorption. Most clinical experience were gained from toxicological poisoning with nickel carbonyl and iron pentacarbonyl. Nickel carbonyl is considered as one of the strongest inhalation poisons.

Inhalation of nickel carbonyl causes acute non-specific symptoms similar to a carbon monoxide poisoning as nausea, cough, headaches, fever and dizziness. After some time, severe pulmonary symptoms such as cough, tachycardiacyanosis or problems in the gastrointestinal tract occur. In addition to pathological alterations of the lung, such as by metallation of the alveoli, damages are observed in the brain, liver, kidneys, adrenal glands and the spleen. A metal carbonyl poisoning often requires a long-lasting recovery.

Chronic exposure by inhalation of low concentrations of nickel carbonyl can cause neurological symptoms such as insomnia, headaches, dizziness and memory loss. Nickel carbonyl is considered carcinogenic, but it can take 20 to 30 years from the start of exposure to the clinical manifestation of cancer.

History

Justus von Liebig (1860)

Initial experiments on the reaction of carbon monoxide with metals were carried out by Justus von Liebig in 1834. By passing carbon monoxide over molten potassium he prepared a substance having the empirical formula KCO, which he called *Kohlenoxidkalium*. As demonstrated later, the compound was not a metal carbonyl, but the potassium salt of ´hexahydroxy benzene and the potassium salt of dihydroxy acetylene

Ludwig Mond, circa 1909

The synthesis of the first true heteroleptic metal carbonyl complex was performed by Paul Schütz-zenberger in 1868 by passing chlorine and carbon monoxide over platinum black, where dicarbonyldichloroplatinum ($Pt(CO)_2Cl_2$) was formed.

Ludwig Mond, one of the founders of Imperial Chemical Industries, investigated in the 1890s with Carl Langer and Friedrich Quincke various processes for the recovery of chlorine which was lost in the Solvay process by nickel metals, oxides and salts. As part of their experiments the group treated nickel with carbon monoxide. They found that the resulting gas colored the gas flame of a burner in a greenish-yellowish color; when heated in a glass tube it formed a nickel mirror. The gas could be condensed to a colorless, water-clear liquid with a boiling point of 43 °C. Thus, Mond and his coworker had discovered the first pure, homoleptic metal carbonyl, nickel tetracarbonyl ($Ni(CO)_4$). The unusual high volatility of the metal compound nickel tetracarbonyl led Kelvin with the statement that Mond had "given wings to the heavy metals".

The following year, Mond and Marcellin Berthelot independently discovered iron pentacarbonyl, which is produced by a similar procedure as nickel tetracarbonyl. Mond recognized the economic potential of this class of compounds, which he commercially used in the Mond process and financed more research on related compounds. Heinrich Hirtz and his colleague M. Dalton Cowap synthesized metal carbonyls of cobalt, molybdenum, ruthenium, and diiron nonacarbonyl. In 1906 James Dewar and H. O. Jones were able to determine the structure of di-iron nonacarbonyl, which is produced from iron pentacarbonyl by the action of sunlight. After Mond, who died in 1909, the chemistry of metal carbonyls fell for several years in oblivion. The BASF started in 1924 the industrial production of iron pentacarbonyl by a process which was developed by Alwin Mittasch. The iron pentacarbonyl was used for the production of high-purity iron, so-called carbonyl iron, and iron oxide pigment. Not until 1927 did A. Job and A. Cassal succeed in the preparation of chromium hexacarbonyl and tungsten hexacarbonyl, the first synthesis of other homoleptic metal carbonyls.

Walter Hieber played in the years following 1928 a decisive role in the development of metal carbonyl chemistry. He systematically investigated and discovered, among other things, the Hieber base reaction, the first known route to metal carbonyl hydrides and synthetic pathways leading to metal carbonyls such as dirhenium decacarbonyl. Hieber, who was since 1934 the Director of the Institute of Inorganic Chemistry at the Technical University Munich published in four decades 249 papers on metal carbonyl chemistry.

Kaiser Wilhelm Institute for Coal Research
(now Max Planck Institute for Coal Research)

Also in the 1930s Walter Reppe, an industrial chemist and later board member of the BASF, discovered a number of homogeneous catalytic processes, such as the hydrocarboxylation, in which olefins or alkynes react with carbon monoxide and water to form products such as unsaturated acids and their derivatives. In these reactions, for example, nickel carbonyl or cobalt carbonyls act as catalysts. Reppe also discovered the cyclotrimerization and tetramerization of acetylene and its derivatives to benzene and benzene derivatives with metal carbonyls as catalysts. BASF built in the 1960s a production facility for acrylic acid by the Reppe process, which was only superseded in 1996 by more modern methods based on the catalytic propylene oxidation.

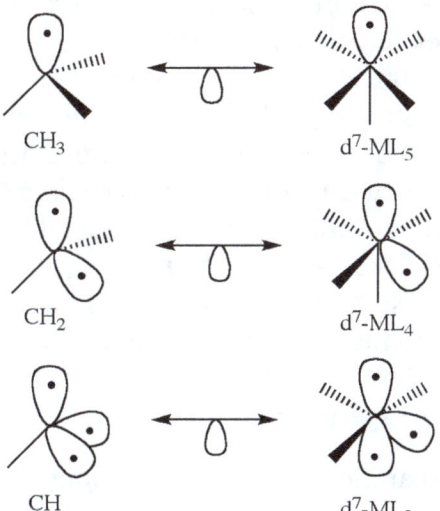

Isolobal fragments with tetrahedral or octahedral geometry

For the rational design of new complexes the concept of the isolobal analogy has been found useful. Roald Hoffmann was awarded with the Nobel Prize in chemistry for the development of the concept. The concept describes metal carbonyl fragments of $M(CO)_n$ as parts of octahedral building blocks in analogy to the tetrahedral CH_3-, CH_2- or CH- fragments in organic chemistry. In example Dimanganese decacarbonyl is formed in terms of the isolobal analogy of two $d^7Mn(CO)_5$ fragments, that are isolobal to the methyl radical $CH_3\cdot$. In analogy to how methyl radicals combine to form Ethane, these can combine to dimanganese decacarbonyl. The presence of isolobal analog fragments does not mean that the desired structures can be synthezied. In his Nobel Prize lecture Hoffmann emphasized that the isolobal analogy is a useful but simple model, and in some cases does not lead to success.

The economic benefits of metal-catalysed carbonylations, e.g. Reppe chemistry and hydroformylation, led to growth of the area. Metal carbonyl compounds were discovered in the active sites of three naturally occurring enzymes.

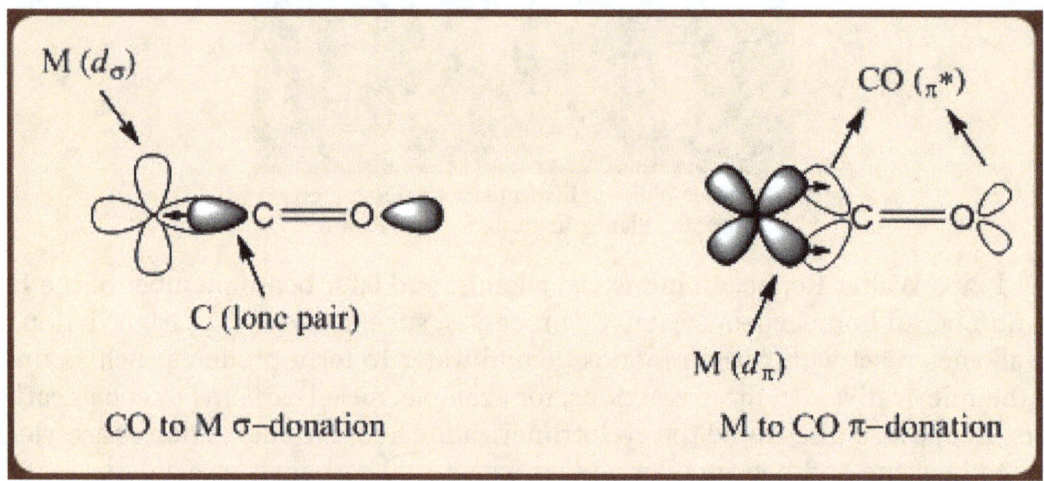

Orbital diagram showing ligand to metal forward σ-donation and the metal
to ligand backward π-donation in metal-CO interaction.

In the metal carbonyl complexes, the direct bearing of the π−back donation is observed on the M−C bond distance that becomes shorter as compared to that of a normal M−C single bond distance. For example, the $CpMo(CO)_3CH_3$ complex, exhibits two kind of M−C bond distances that comprise of a longer $Mo-CH_3$ distance (2.38 Å) and a much shorter Mo−CO distance (1.99 Å) arising out of a metal to ligand π−back donation. It becomes thus apparent that the metal−CO interaction can be easily characterized using X−ray crystallography. The infrared spectroscopy can also be equally successfully employed in studying the metal−CO interaction. Since the metal to CO π−back bonding involves a π−donation from the metal d_π orbital to a π* orbital of a C−O bond, significant shift of the ν(CO) stretching frequency towards the lower energy is observed in metal carbonyl complexes with respect to that of free CO (2143 cm^{-1}).

Preparation of Metal Carbonyl Complexes

The common methods of the preparation of the metal carbonyl compounds are,

i.   *Directly using CO*

$$Fe \xrightarrow{CO, 200\,atm, 200^\circ C} Fe(CO)_5$$

The main requirement of this method is that the metal center must be in a reduced low oxidation state in order to facilitate CO binding to the metal center through metal to ligand π−back donation.

ii. Using CO and a reducing agent

$$NiSO_4 + CO + S_2O_4^{2-} \rightarrow Ni(CO)_4$$

This method is commonly called reductive carbonylation and is mainly used for the compounds having higher oxidation state metal centers. The reducing agent first reduces the metal center to a lower oxidation state prior to the binding of CO to form the metal carbonyl compounds.

iii. From carbonyl compounds

This method involves abstraction of CO from organic compounds like the alcohols, aldehydes and CO_2.

Reactivities of metal carbonyls

i. Nucleophilic attack on carbon

The reaction usually gives rise to carbene moiety.

ii. Electrophilic attack at oxygen

$$Cl(PR_3)_4\,Re\text{-}CO + AlMe_3 \rightarrow \left[Cl(PR_3)_4\,Re\text{-}CO \rightarrow AlMe_3\right]$$

iii. Migratory insertion reaction

$$MeMn(CO)_5 + PMe_3 \rightarrow (MeCO)Mn(CO)_4(PMe_3)$$

The metal carbonyl displays two kinds of bindings in the form of the terminal and the bridging modes. The infrared spectroscopy can easily distinguish between these two binding modes of the metal carbonyl moiety as the terminal ones show v(CO) stretching band at ca. 2100-2000 cm^{-1} while the bridging ones appear in the range 1720–1850 cm^{-1}. The carbonyl moiety can bridge between more than two metal centers.

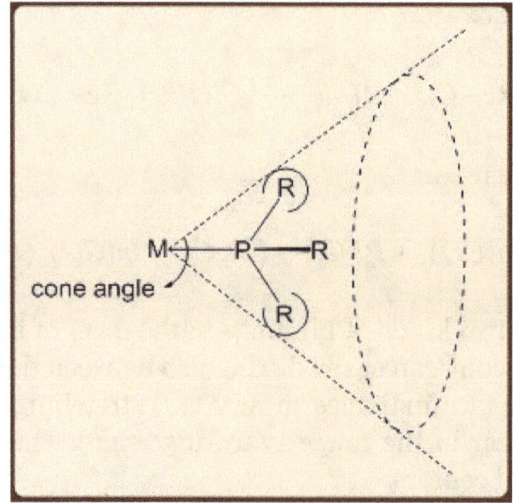

Different bridging modes of the carbonyl binding to a metal is shown.

Metal Phosphines

Phosphines are one of the few ligands that have been extensively studied over the last few decades to an extent that the systematic fine tuning of the sterics and electronics can now be achieved with certain degree of predictability. Phosphines are better spectator ligands than actor ligands. Tolman carried out pioneering infrared spectroscopy experiments on the $PR_3Ni(CO)_3$ complexes looking at the $v_{(CO)}$ stretching frequencies for obtaining an insight on the donor properties of the PR_3 ligands. Thus, a stronger σ–donor phosphine ligand would increase the electron density at the metal center leading to an enhanced metal to ligand π–back bonding and thereby lowering of the $v_{(CO)}$ stretching frequencies in these complexes. Another important aspect of the phosphine ligand is its size that has significant steric impact on its metal complexes. Thus, unlike CO ligand, which is small and hence many may simultaneously be able to bind to a metal center, the same is not true for the phosphine ligands as only a few can bind to a metal center. The number of phosphine ligands that can bind to a metal center also depends on the size of its R substituents. For example, up to two can bind to a metal center in case of the PCy_3 or $P(i–Pr)_3$ ligands, three or four for PPh_3, four for Me_2PH, and five or six for PMe_3. The steric effect of phosphine was quantified by Tolmann and is given by a parameter called Cone Angle that measures the angle at the metal formed by the PR_3 ligand binding to a metal.

Cone Angle in metal–phosphine complexes.

The *Cone Angle* criteria has been successfully invoked in rationalizing the properties of a wide range of metal phosphine complexes. One unique feature of the phosphine ligand is that it allows convenient change of electronic effect without undergoing much change in its steric effects. For example, PBu_3 and $P(O^iPr)_3$ have similar steric effects but vary in their electronic effects. The converse is also true as the steric effect can be easily changed without undergoing much change in the electronic effect. For example, PMe_3 and $P(o-tolyl)_3$ have similar electronic effect but differ in their steric effects. Thus, the ability to conveniently modulate the steric and the electronic effects make the phosphine ligands a versatile system for carrying out many organometallic catalysis.

Structure and Bonding

Phosphines are two electron donors that engage a lone pair for binding to metals. These are thus considered as good σ–donors and poor π–acceptors and they belong to the same class with the aryl, dialkylamino and alkoxo ligands. In fact they are more π–acidic than pure σ–donor ligands like NH_3 and, more interestingly so, their π–acidity can be varied significantly by systematic incorporation of substituents on the P atom. For example, PF_3 is more π–acidic than CO. Analogous to what is observed in case of the benchmark π–acidic CO ligand, in which the metal d_π orbital donates electron to a π* orbital of a C–O bond, in the case of the phosphines ligands, such π–back donation occurs from the metal d_π orbital occurs on to a σ* orbital of a P–R bond. In phosphine ligands, with the increase of the electronegativity of R both of the σ and the σ* orbitals of the P–R bond gets stabilized. Consequently, the contribution of the atomic orbital of the P atom to the σ*–orbital of the P–R bond increases, which eventually increases the size of the σ* orbital of the P–R bond. This in turn facilitates better overlap of the σ* orbital of the P–R bond with the metal d_π orbital during the metal to ligand π–back donation in these metal phosphine complexes.

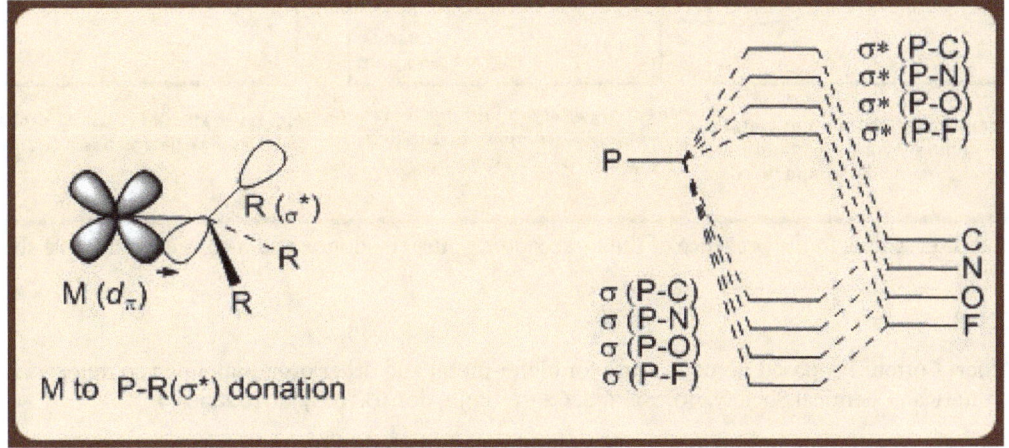

Back donation from the metal dπ orbital to a σ* orbital of a P–R bond.

Starting from CO, which is a strong π–acceptor ligand, to moving to the phosphines, which are good σ–donors and poor π–acceptor ligands, to even going further to other extreme to the ligands, which are both good σ–donors as well as π–donors, a rich variety of phosphine ligands thus are available for stabilizing different types of organometallic complexes. In this context the following ligands are discussed below.

π-basic Ligands

Alkoxides (RO⁻) and halides like F⁻, Cl⁻ and Br⁻ belong to a category of π–basic ligands as they engage a second lone pair for π–donation to the metal over and above the first lone pair partaking σ–donation to the metal. Opposite to what is observed in the case of π–acidic ligands, in which the π* ligand orbital stabilizes the d_π metal orbital and thereby affecting a larger ligand field splitting, as consistent with the strong field nature of these ligands, in the case of the π–basic ligands, the second lone pair destabilizes the d_π metal orbitals leading to a smaller ligand field splitting, which is in agreement with the weak field nature of these ligands. The orbitals containing the lone pair of the ligands are usually located on the more electronegative heteroatoms and so they are invariably lower in energy than the metal d_π orbitals. Hence, the destabilization of the metal d_π orbitals occurs due to the repulsion of the filled ligand lone pair orbital with the filled metal d_π orbitals. In case of the situations in which the metal d_π orbitals are vacant, like in d^o systems of Ti⁴⁺ ions, the possibility of the destabilization of the metal d_π orbitals do not arise but instead stabilization occurs through the donation of the filled ligand lone pair orbital electrons to the empty metal d_π orbitals as seen in the case of TiF₆ and W(OMe)₆. Thus, this scenario in π–basic ligands is opposite to that observed in case of the π–acidic ligands, for which the empty π* ligand orbitals are higher in energy than the filled metal d_π orbitals.

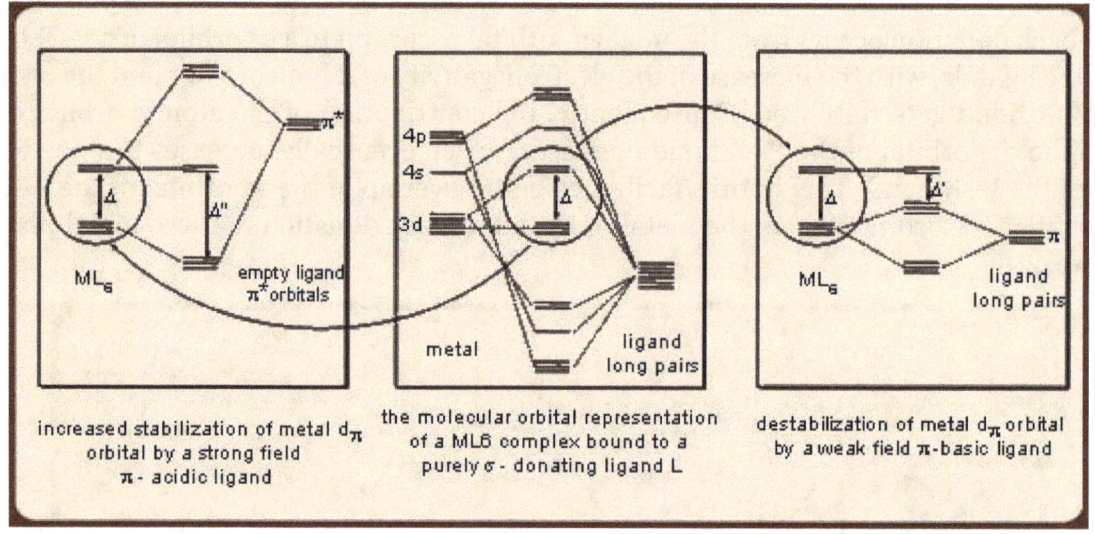

Orbital interactions in the presence of the π–acceptor, (pure) σ–donor and π–basic ligands are shown.

References

- F. Albert Cotton: Proposed nomenclature for olefin-metal and other organometallic complexes. In: Journal of the American Chemical Society. 90, 1968, S. 6230–6232, doi:10.1021/ja01024a059

- Dyson, P. J.; McIndoe, J. S. (2000). Transition Metal Carbonyl Cluster Chemistry. Amsterdam: Gordon & Breach. ISBN 90-5699-289-9

- Crabtree, R. H. (2005). "4. Carbonyls, Phosphine Complexes, and Ligand Substitution Reactions". The Organometallic Chemistry of the Transition Metals (4th ed.). pp. 87–124. doi:10.1002/0471718769

- Sargent, A. L.; Hall, M. B. (1989). "Linear Semibridging Carbonyls. 2. Heterobimetallic Complexes Containing a Coordinatively Unsaturated Late Transition Metal Center". Journal of the American Chemical Society. 111 (5): 1563–1569. doi:10.1021/ja00187a005

- Spessard, G. O.; Miessler, G. L. (2010). Organometallic Chemistry (2nd ed.). New York: Oxford University Press. pp. 79–82. ISBN 978-0-19-533099-1

- Mittasch, A. (1928). "Über Eisencarbonyl und Carbonyleisen". Angewandte Chemie. 41 (30): 827–833. doi:10.1002/ange.19280413002

- Li, P.; Curtis, M. D. (1989). "A New Coordination Mode for Carbon Monoxide. Synthesis and Structure of $Cp_4Mo_2Ni_2S_2(\eta^1, \mu_4\text{-CO})$". Journal of the American Chemical Society. 111 (21): 8279–8280. doi:10.1021/ja00203a040

- Holleman, A. F.; Wiberg, E.; Wiberg, N. (2007). Lehrbuch der Anorganischen Chemie (102nd ed.). Berlin: de Gruyter. pp. 1780–1822. ISBN 978-3-11-017770-1

- Hieber, W.; Fuchs, H. (1941). "Über Metallcarbonyle. XXXVIII. Über Rheniumpentacarbonyl". Zeitschrift für anorganische und allgemeine Chemie. 248 (3): 256–268. doi:10.1002/zaac.19412480304

- Ohst, H. H.; Kochi, J. K. (1986). "Electron-Transfer Catalysis of Ligand Substitution in Triiron Clusters". Journal of the American Chemical Society. 108 (11): 2897–2908. doi:10.1021/ja00271a019

- Harris, D. C.; Bertolucci, M. D. (1980). Symmetry and Spectroscopy: Introduction to Vibrational and Electronic Spectroscopy. Oxford University Press. ISBN 978-0-19-855152-2

- Pearson, R. G. (1995). "The Transition-Metal-Hydrogen Bond". Chemical Reviews. 85 (1): 41–49. doi:10.1021/cr00065a002

- Trout, W. E. Jr. (1937). "The Metal Carbonyls. I. History; II. Preparation". Journal of Chemical Education. 14 (10): 453. Bibcode:1937JChEd..14..453T. doi:10.1021/ed014p453

- Fairweather-Tait, S. J.; Teucher, B. (2002). "Iron and Calcium Bioavailability of Fortified Foods and Dietary Supplements". Nutrition Reviews. 60 (11): 360–367. doi:10.1301/00296640260385801

- Henderson, W.; McIndoe, J. S. Mass Spectrometry of Inorganic, Coordination and Organometallic Compounds: Tools – Techniques – Tips. John Wiley & Sons. ISBN 0-470-85015-9

- Mond, L.; Langer, C.; Quincke, F. (1890). "Action of Carbon Monoxide on Nickel". Journal of the Chemical Society, Transactions. 57: 749–753. doi:10.1039/CT8905700749

7

Oxidative Addition and Insertion Reaction

Through oxidative addition, two anionic ligands and added to the metal centre and its coordination number and oxidation state is increased. Reductive elimination is the opposite of oxidative addition, through which two ligands are removed from the metal centre. It forms a covalent bond between ligands and leads to the decrease in the oxidation state the metal centre. The section also deals with insertion and elimination reaction, and nucleophilic and electrophilic addition and abstraction. All the diverse principles of organometallic chemistry have been carefully analyzed in this chapter.

Oxidative Addition and Reductive Elimination

Oxidative addition (OA) is a process that adds two anionic ligands *e. g.* A and B, that originally are a part of a A-B molecule, like in H_2 or Me−I, on to a metal center and is of significant importance from the perspective of both synthesis and catalysis. The exact reverse of the same process, in which the two ligands, A and B, are eliminated from the metal center forming back the A−B molecule, is called the reductive elimination (RA). As A and B are anionic X type ligands, the oxidative addition is accompanied by an increase in the coordination number, valence electron count as well as in the formal oxidation state of the metal center by two units. The oxidative addition step may proceed by a variety of pathways. It requires the metal center to be both coordinatively unsaturated and electron deficient.

$$L_nM \; + \; A{-}B \; \underset{RE}{\overset{OA}{\rightleftharpoons}} \; L_nM\overset{\diagup A}{\underset{\diagdown B}{}}$$

$$\text{16 VE} \qquad\qquad\qquad \text{18 VE}$$

$$\Delta \text{O.S.} = +2$$
$$\Delta \text{C.N.} = +2$$

Oxidative addition transfers a single mononuclear metal center having 16 VE to a 18 VE species upon oxidative addition. Another frequently observed pathway is that a 18 VE complex looses a ligand to become a 16 VE species which then undergoes an oxidative addition. Apart from above two types, another possible pathway for oxidative addition proceeds as a binuclear oxidative addition in which each of the two metal centers undergo change in oxidation state, electron count and coordination number by one unit instead of two. This type of a binuclear oxidative addition is observed for a 17 VE metal complex or for a binuclear 18 VE metal complex having a metal–metal bond and, for which the metal has a stable oxidation state at a higher positive oxidation state by one unit.

$$2L_nM \quad or \quad L_nM - ML_n \underline{A - B} \; L_nM - A + L_nM - B$$

$$17\,VE \qquad 18\,VE \qquad\qquad 18VE \qquad 18VE$$

$$\Delta O.S. = +1 \quad \Delta O.S. = +1$$
$$\Delta C.N. = +1 \quad \Delta C.N. = +1$$

It is interesting to note that in the oxidative addition the breakage of A–B σ–bond occurs as a result of a net transfer of electrons from the metal center to a σ*–orbital of the A–B bond, thus resulting in the formation of the two new M–A and M–B bonds. The oxidative addition is facilitated by electron rich metal centers having low oxidation state whereas the reductive elimination is facilitated by metal centers in higher oxidation state.

Table 1. Common types of oxidative addition reactions.

Change in d^n Configuration	Change in Coordination Geometry	Examples	Group
$d^{10} \rightarrow d^8$	Lin. $\xrightarrow{X_2}$ Sq. Pl.	Au(I) $\rightarrow$ (III)	11
	Tet. $\xrightarrow{-2L, X_2}$ Sq. Pl.	Pt, Pd(0) $\rightarrow$ (II)	10
$d^8 \rightarrow d^6$	Sq. Pl. $\xrightarrow{X_2}$ Oct.	Pt, Pd(II) $\rightarrow$ (IV)	10
		Rh, Ir(I) $\rightarrow$ (III)	9
		Pt, Pd(0) $\rightarrow$ (II)	8
	TBP. $\xrightarrow{-L, X_2}$ Oct.	Pt, Pd(I) $\rightarrow$ (III)	9
		Pt, Pd(0) $\rightarrow$ (II)	8
$d^7 \rightarrow d^6$	2Sq. Pyr. $\xrightarrow{X_2}$ 2Oct.	2Co(II) $\rightarrow$ (III)	8
	2Oct. $\xrightarrow{-L, X_2}$ 2Oct.	2Co(II) $\rightarrow$ (III)	8
$d^6 \rightarrow d^4$	Oct. $\xrightarrow{X_2}$ 7-c	Re(I) $\rightarrow$ (III)	7
		Pt, Pd(0) $\rightarrow$ (II)	6
		V(-I) $\rightarrow$ (I)	5
$d^4 \rightarrow d^3$	2Sq. Pyr. $\xrightarrow{X_2}$ 2Oct.	2Cr(II) $\rightarrow$ (III)	6
	2Oct. $\xrightarrow{-L, X_2}$ 2Oct.	2Cr(II) $\rightarrow$ (III)	6
$d^4 \rightarrow d^2$	Oct. $\xrightarrow{X_2}$ 8-c	Mo, W(II) $\rightarrow$ (IV)	6
$d^2 \rightarrow d^0$	Various	Pt, Pd(III) $\rightarrow$ (V)	5
		Pt, Pd(II) $\rightarrow$ (IV)	4

Abbreviations: Lin. = linear, Tet. = tetrahedral, Oct. = octahedral, Sq. Pl. = square planar, TBP = trigonal bipyramidal, Sq. Pyr. = square pyramidal: 7-c, 8-c = 7- and 8-coordinate.

In principle, the oxidative addition is the reverse of reductive elimination, but in practice one may dominate over the other. Thus, the favorability of one over the other is depends on the position of equilibrium, which is further dependent on the stability of the two oxidation states of the metal and on the difference of bond strengths of A−B versus that of the M−A and M−B bonds. For example, metal hydride complexes frequently undergo reductive elimination to give alkanes but rarely an alkane undergoes oxidative addition to give an alkyl hydride complex. Along the same line, alkyl halides frequently undergo oxidative addition to a metal giving metal−alkyl halide complexes but these complexes rarely reductively eliminate to give back alkyl halides. Usually the oxidative addition is more common for 3rd row transition metals because they tend to possess stronger metal ligand bond strengths. The oxidative addition is also favored by strong donor ligands, as they stabilize the higher oxidation state of the metal. The oxidative addition reaction can expand beyond transition metals as observed in the case of the Grignard reagents as well as for some main group elements.

Oxidative addition may proceed by several pathways as discussed below.

Concerted Oxidative Addition Pathway

Oxidative addition may proceed by a concerted 3−centered associative mechanism involving the incoming ligand with the metal center. Specifically, the addition proceeds by the formation of a σ−complex upon binding of an incoming ligand say, H$_2$, followed by the cleavage of the H−H bond as a result of the back donation of electrons from the metal to the σ*−orbital of the H−H bond. Such type of addition is common for the H−H, C−H and Si−H bonds. As expected these proceed by two steps (*i*) the formation of a σ−complex and (*ii*) the oxidation step. For example, the oxidative addition of H$_2$ to Vaska's complex (PMe$_3$)$_2$Ir(CO)Cl proceeds by this pathways.

S$_N$2 pathway

This pathway of oxidative addition is operational for the polarized AB type of ligand substrates like the alkyl, acyl, allyl and benzyl halides. In this mechanism, the L$_n$M fragment directly donates electrons to the σ*−orbital of the A−B bond by attacking the least electronegative atom, say A, of the AB molecule and concurrently initiating the elimination of the most electronegative atom of the AB molecule in its anionic form, B$^-$. These reactions proceed *via* a polar transition state that is accompanied by an inversion of the stereochemistry at the atom of attack by the metal center and are usually accelerated in polar solvents.

Radical Pathway

This type of oxidative addition proceeds *via* a by radical pathway that generally are vulnerable to the presence of impurities. The radical processes can be of non–chain and chain types. In a non–chain type of mechanism, the metal (M) transfer one electron to the σ*–orbital of the RX bond resulting in the formation of a radical cation $M^{+\cdot}$ and a radical anion $RX^{-\cdot}$. The generation of the two radical fragments occurs by the way of the elimination of the anion X^- from the radical anion $RX^{-\cdot}$ leaving behind the radical $R^\cdot$ while the subsequent reaction of X^- anion with the radical cation $M^{+\cdot}$ generates the other radical $MX^\cdot$ in the course of the reaction. Such type of non–chain type of oxidative addition is observed for the addition of the alkyl halide to $Pt(PPh_3)_3$ complexes.

$$PtL_3 \xrightarrow{\text{fast}} PtL_2$$

$$PtL_2 + RX \rightarrow {}^\cdot PtL_2 + {}^{-\cdot}RX \xrightarrow{\text{slow}} {}^\cdot PtXL_2 + {}^\cdot R$$

$${}^\cdot PtXL_2 + {}^\cdot R \xrightarrow{\text{fast}} RPtXL_2$$

The other type in this category is the chain radical type reaction that is usually observed for the oxidative addition of EtBr and $PhCH_2Br$ to the $(PMe_3)_2Ir(CO)Cl$ complex. For this process a radical initiator is required and the reaction proceeds along a series of known steps common to a radical process.

$$R^\cdot + Ir^\cdot Cl(CO)L_2 \rightarrow RIr^{\cdot\cdot} Cl(CO)L_2$$

$$RIr^{\cdot\cdot} Cl(CO)L_2 + RX \rightarrow RXIr^{\cdot\cdot} Cl(CO)L_2 + R^\cdot$$

$$2R^\cdot \rightarrow R_2$$

Ionic Pathway

This is kind of pathway for the oxidative addition reaction is common to the addition of hydrogen halides (HX) in its dissociated H^+ and X^- forms. The ionic pathways are usually of the following two types (i) the ones in which the starting metal complex adds to H^+ prior to the addition of the halide X^- and (ii) the other type, in which the halide anion X^- adds to the starting metal complex first, and then the addition of proton H^+ occurs on the metal complex.

$$Pt(PPh_3)_4 + H^+ + Cl^- \xrightarrow{-PPH_3} \left[HPt(PPh_3)_3 \right]^+ + Cl^- \xrightarrow{-PPH_3} \left[HPt(PPh_3)_2 \right]$$

$18VE, d^{10}$	$16VE, d^8$	$16VE, d^8$
tetrahedral	*square planar*	*square planar*

$$\left[Ir(cod)L_2 \right]^+ + Cl^- + H^+ \rightarrow \left[Ir(cod)L_2 \right] + H^+ \rightarrow \left[IrHCl(cod)L_2 \right]^+$$

$16VE, d^8$	$18VE, d^8$	$18VE, d^6$
tetrahedral	*TBP*	*TBP*

Reductive Elimination

The reductive eliminations are reverse of the oxidative addition reactions and are accompanied by the reduction of the formal oxidation state of the metal and the coordination numbers by two units. The reductive eliminations are commonly observed for d^8 systems, like the Ni(II), Pd(II) and Au(III) ions and the d^6 systems, like the Pt(IV), Pd(IV), Ir(III) and Rh(III) ions. The reaction may proceed by the elimination of several groups.

$$L_n MRH \rightarrow L_n M + R - H$$
$$L_n MR_2 \rightarrow L_n M + R - R$$
$$L_n MH(COR) \rightarrow L_n M + RCHO$$
$$L_n MR(COR) \rightarrow L_n M + R_2 CO$$
$$L_n MR(SiR_3) \rightarrow L_n M + R - SiR_3$$

Binuclear Reductive Elimination

Similar to what has been observed in the case of binuclear oxidative addition, the binuclear reductive elimination is also observed in some instances. As expected, the oxidation state and the coordination number decrease by one unit in the binuclear reductive elimination pathway.

$$2MeCH = CHCu(PBu_3) \xrightarrow{heat} MeCh = CHCH = CHMe$$
$$ArCOMn(CO)_5 + HMn(CO)_5 \rightarrow ArCHO + Mn_2(CO)_{10}$$

Oxidative Addition

Oxidative addition and reductive elimination are two important and related classes of reactions in organometallic chemistry. Oxidative addition is a process that increases both the oxidation state and coordination number of a metal centre. Oxidative addition is often a step in catalytic cycles, in conjunction with its reverse reaction, reductive elimination.

Role in transition metal chemistry

For transition metals, oxidative reaction results in the decrease in the d^n to a configuration with fewer electrons, often 2e fewer. Oxidative addition is favored for metals that are (i) basic and/or (ii) easily oxidized. Metals with a relatively low oxidation state often satisfy one of these requirements, but even high oxidation state metals undergo oxidative addition, as illustrated by the oxidation of Pt(II) with chlorine:

$$[PtCl_4]^{2-} + Cl_2 \rightarrow [PtCl_6]^{2-}$$

In classical organometallic chemistry, the formal oxidation state of the metal and the electron count of the complex both increase by two. One-electron changes are also possible and in fact some oxidative addition reactions proceed via series of 1e changes. Although oxidative additions can occur with the insertion of a metal into many different substrates, oxidative additions are most commonly seen with H–H, H–X, and C–X bonds because these substrates are most relevant to commercial applications.

Oxidative addition requires that the metal complex have a vacant coordination site. For this reason, oxidative additions are common for four- and five-coordinate complexes.

Reductive elimination is the reverse of oxidative addition. Reductive elimination is favored when the newly formed X–Y bond is strong. For reductive elimination to occur the two groups (X and Y) should be mutually adjacent on the metal's coordination sphere. Reductive elimination is the key product-releasing step of several reactions that form C–H and C–C bonds.

Mechanisms of Oxidative Addition

Oxidative additions proceed via many pathways that depend on the metal center and the substrates.

Concerted Pathway

Oxidative additions of nonpolar substrates such as hydrogen and hydrocarbons appear to proceed via concerted pathways. Such substrates lack π-bonds, consequently a three-centered σ complex is invoked, followed by intramolecular ligand bond cleavage of the ligand (probably by donation of electron pair into the sigma* orbital of the inter ligand bond) to form the oxidized complex. The resulting ligands will be mutually *cis*, although subsequent isomerization may occur.

$$L_nM \ + \ A—B \ \longrightarrow \ L_nM—\underset{B}{\overset{A}{|}} \ \longrightarrow \ L_nM\overset{A}{\underset{B}{<}}$$

This mechanism applies to the addition of homonuclear diatomic molecules such as H_2. Many C–H activation reactions also follow a concerted mechanism through the formation of an M–(C–H) agostic complex.

A representative example is the reaction of hydrogen with Vaska's complex, *trans*-IrCl(CO) $[P(C_6H_5)_3]_2$. In this transformation, iridium changes its formal oxidation state from +1 to +3. The product is formally bound to three anions: one chloride and two hydride ligands. As shown below, the initial metal complex has 16 valence electrons and a coordination number of four whereas the product is a six-coordinate 18 electron complex.

Ir(I), 16 e   Ir(III), 18 e

Formation of a trigonal bipyramidal dihydrogen intermediate is followed by cleavage of the H–H bond, due to electron back donation into the H–H σ*-orbital. This system is also in chemical equilibrium, with the reverse reaction proceeding by the elimination of hydrogen gas with simultaneous reduction of the metal center.

The electron back donation into the H–H σ*-orbital to cleave the H–H bond causes electron-rich metals to favor this reaction. The concerted mechanism produces a *cis* dihydride, while the stereochemistry of the other oxidative addition pathways do not usually produce *cis* adducts.

SN2-type

Some oxidative additions proceed analogously to the well known bimolecular nucleophilic substitution reactions in organic chemistry. Nucleophillic attack by the metal center at the less electronegative atom in the substrate leads to cleavage of the R–X bond, to form an $[M–R]^+$ species. This step is followed by rapid coordination of the anion to the cationic metal center. For example, reaction of a square planar complex with methyl iodide:

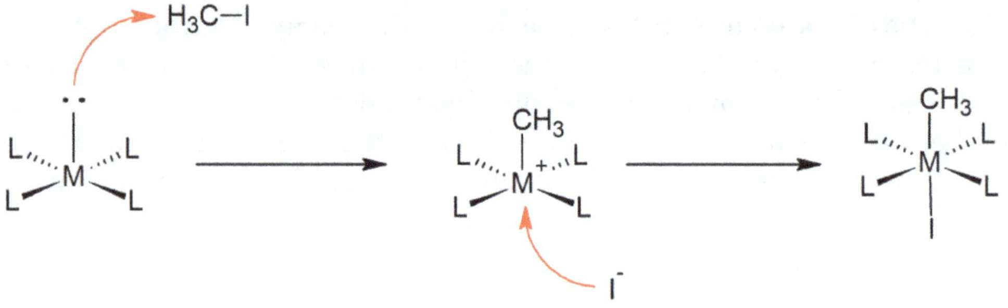

This mechanism is often assumed in the addition of polar and electrophilic substrates, such as alkyl halides and halogens.

Ionic

The ionic mechanism of oxidative addition is similar to the SN$_2$ type in that it involves the stepwise addition of two distinct ligand fragments. The key difference being that ionic mechanisms involve substrates which are dissociated in solution prior to any interactions with the metal center. An example of ionic oxidative addition is the addition of hydrochloric acid.

Radical

In addition to undergoing S$_N$2-type reactions, alkyl halides and similar substrates can add to a metal center via a radical mechanism, although some details remain controversial. Reactions which are generally accepted to proceed by a radical mechanism are known however. One example was proposed by Lednor and co-workers.

Initiation

$$[(CH_3)_2C(CN)N]_2 \rightarrow 2\,(CH_3)_2(CN)C^\bullet + N_2$$

$$(CH_3)_2(CN)C^\bullet + PhBr \rightarrow (CH_3)_2(CN)CBr + Ph^\bullet$$

Propagation

$$Ph^\bullet + [Pt(PPh_3)_2] \rightarrow [Pt(PPh_3)_2Ph]^\bullet$$

$$[Pt(PPh_3)_2Ph]^\bullet + PhBr \rightarrow [Pt(PPh_3)_2PhBr] + Ph^\bullet$$

Applications

Oxidative addition and reductive elimination are invoked in many catalytic processes both in homogeneous catalysis (i.e., in solution) such as the Monsanto process and alkenehydrogenation using Wilkinson's catalyst. It is often suggested that oxidative addition-like reactions are also involved in mechanisms of heterogeneous catalysis, e.g. hydrogenation catalyzed by platinum metal. Metals are however characterised by band structures, so oxidation states are not meaningful. Oxidative addition is also needed in order for nucleophilic addition of an alkyl group to occur. Oxidative insertion is also a crucial step in many cross-coupling reactions like the Suzuki coupling, Negishi coupling, and the Sonogashira coupling.

Insertion and Elimination Reaction

Unlike what we have learned about the oxidative addition and the reductive elimination reactions, that facilitate the addition or removal of 1–electron and 2–electron ligands on to a metal center, the insertion and the elimination reactions perform the subsequent transformation of these ligands from within the same coordination sphere of a metal. Thus, in an insertion reaction a metal bound 2 electron A=B type of a ligand can insert on to a M–X bond resulting in a new metal bound 1–electron ligand like, M–A–B–X, which is formed as a result of the formations of the M–A and B–X bonds. The insertion reaction thus leads to the generation of one vacant site created at the initial metal bound A=B site. Thus a primary requirement for reverse elimination reaction to occur is the presence of a *cis* vacant site.

Insertions are of two types, 1,1–insertion and 1,2–insertion. In 1,1–insertion both the metal M and the ligand X of the M–X bond end up on the same atom like in the M–A(X)–B moiety formed after the insertion of the A=B molecule in the M–X bond, whereas in the 1,2–insertion, these end up on the adjacent atoms like in the M–A–B–X moiety formed after the insertion of the A=B molecule in the M–X bond. The type of insertion depends on the type of the ligand undergoing the insertion like η^1–ligand showing 1,1–insertion and η^2–ligands showing 1,2–insertion. For example, the CO ligand exclusively undergoes 1,1–insertion while the C_2H_4 ligand undergoes 1,2–insertion. The SO_2 ligand remains the only exception as it can bind by both η^1–(by S–donor site) and η^2–(by S– and O–donor sites) modes and thus shows both type of insertions.

Though the insertion and the elimination reactions are mutually reversible, owing to the thermo-dynamical reasons one is favored over the other. For example, SO_2 is known to insert into the M−R bond with no report of its elimination is known of it, while for N_2 ligand, no report of its insertion is known but it's elimination from a M−N=N−R bond is known.

$$M - R + SO_2 \rightarrow M - SO_2R$$
$$M - N = N - Ar \rightarrow M - Ar + N_2$$

The CO ligand inserts readily into a metal−alkyl bond. Sterically demanding substituents (R) is found to accelerate the reaction as a bulky R group in an acyl moiety in the final M−CO−R bond is far removed from the metal center than that in the starting M−R bond.

Isonitriles readily insert into M−R and M−H bonds giving η^2-bound iminoacyls.

Olefins usually inserts across a M−H bond and such insertions are of relevance to the commer-cially important olefin polymerization process. In certain cases the 1,2−insertions of olefins give species exhibiting agostic insertions.

ß-Elimination

β−elimination is just a reverse of 1,2−insertion and is a major cause of decomposition of metal alkyl bond having a b−hydrogen atom. A pre-requisite for the β−elimination reaction to occur is the presence of an adjacent vacant site next to the metal alkyl bond undergoing the β−elimination. The β−elimination step results in the formation of a metal hydride species that also contain a metal bound olefin moiety.

α-elimination

In absence of a β–hydrogen, a metal bound alkyl moiety may undergo the cleavage of a C–H bond at the α, γ and δ positions. For example, a methyl moiety may α–eliminate to give a metal bound methylene hydride moiety.

Nucleophilic and Electrophilic Addition and Abstraction

The nucleophilic and electrophilic substitution and abstraction reactions can be viewed as ways of activation of substrates to allow an external reagent to directly attack the metal activated ligand without requiring prior binding of the external reagent to the metal. The attacking reagent may be a nucleophile or an electrophile. The nucleophilic attack of the external reagent is favored if the L_nM fragment is a poor π–base and a good σ–acid *i.e.*, when the complex is cationic and/or when the other metal bound ligands are electron withdrawing such that the ligand getting activated gets depleted of electron density and can undergo an external attack by a nucleophile Nu^-, like LiMe or OH^-. The attack of the nucleophiles may result in the formation of a bond between the nucleophiles and the activated unsaturated substrate, in which case it is called nucleophilic addition, or may result in an abstraction of a part or the whole of the activated ligand, in which case it is called the nucleophilic abstraction. The nucleophilic addition and the abstraction reactions are discussed below.

Nucleophilic Addition

An example of a nucleophilic addition reaction is shown below.

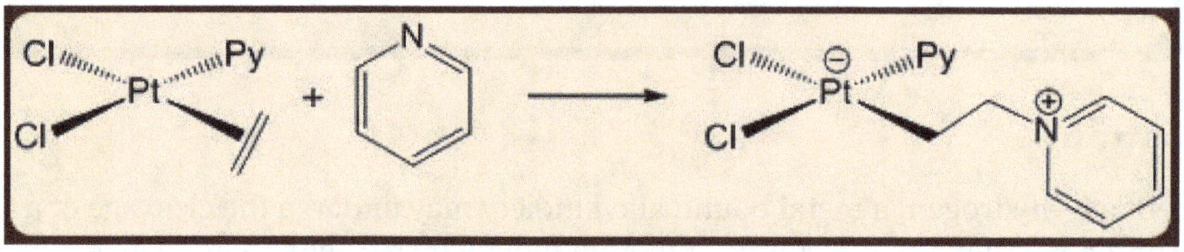

Carbon monoxide (CO) as a ligand can undergo nucleophilic attack when bound to a metal center of poor π–basicity, as the carbon center of the CO ligand is electron deficient owing to the ligand to metal σ–donation not being fully compensated by the metal to ligand π–back donation. Thus, activated CO ligand undergoes nucleophilic attack by the lithium reagent to give an anionic acyl ligand, which upon alkylation generates the famous Fischer carbene complex.

$$Fe(CO)_5 + LiNEt_2 \longrightarrow (CO)_4Fe{=}C(NEt_2)(OLi) + Me_3O^+ \longrightarrow (CO)_4Fe{=}C(NEt_2)(OMe)$$

Nucleophilic Abstraction

An example of a nucleophilic abstraction reaction is shown below.

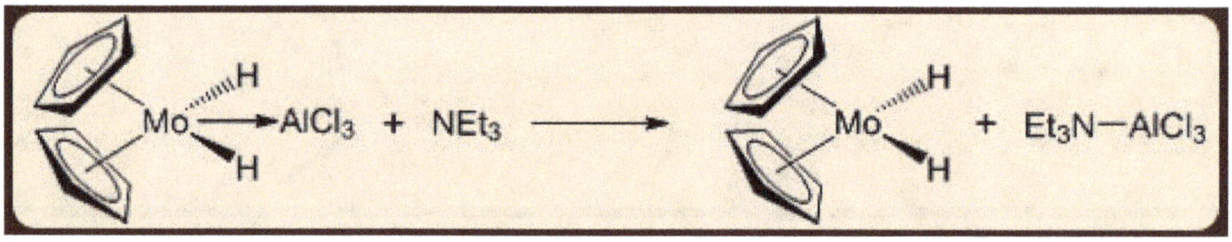

Electrophilic Addition

Similar to the nucleophilic addition and abstraction reactions, the electrophilic counterparts of these reactions also exist. An electrophilic attack is favored if the L_nM fragment is a good π–base and a poor σ–acid *i.e.,* when the complex is anionic with the metal center at low–oxidation state and/or when the other metal bound ligands are electron donating such that the ligand getting activated becomes electron rich from the π–back donation of the metal center and thus can undergo an external attack by an electrophile E^+ like H^+ and CH_3I. The attack of the electrophiles may result in the formation of a bond between the electrophile and the activated unsaturated substrate, in which case it is called electrophilic addition, or may result in an abstraction of a part or the whole of the activated ligand, in which case it is called the electrophilic abstraction.

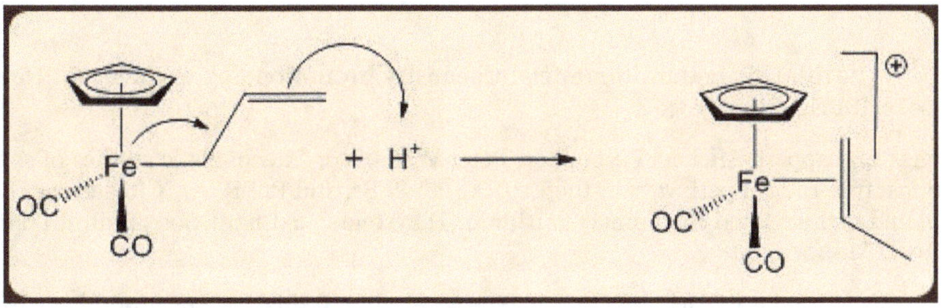

Electrophilic Abstraction

An example of an electrophilic abstraction reaction is shown below.

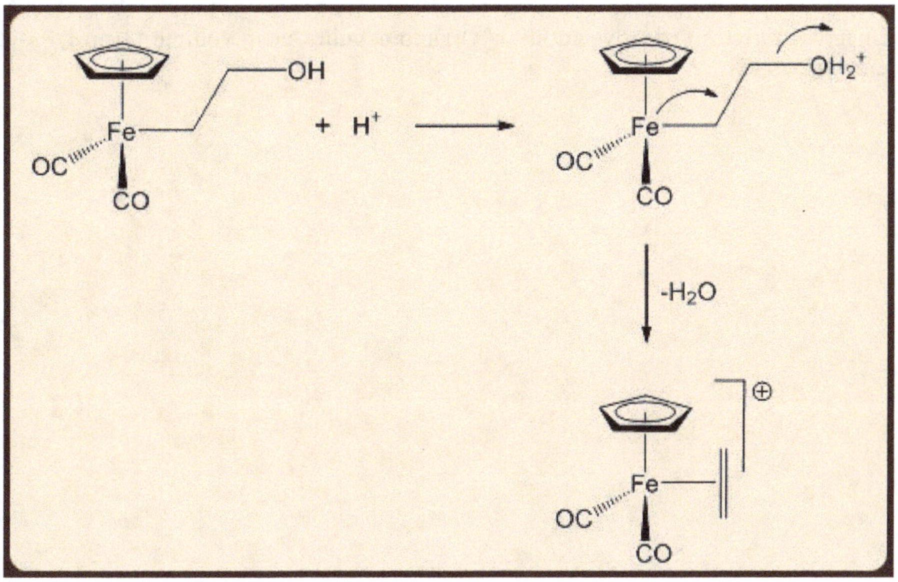

Alkyl abstractions are often achieved by Hg^{2+} that can proceed in two ways, (*i*) by an attack at the α–carbon of a metal alkyl bond leading to an inversion of configuration at the alkyl carbon and (*ii*) by an attack at the metal center leading to retention of configuration at the alkyl carbon. The inversion of configuration proceeds by the following pathway.

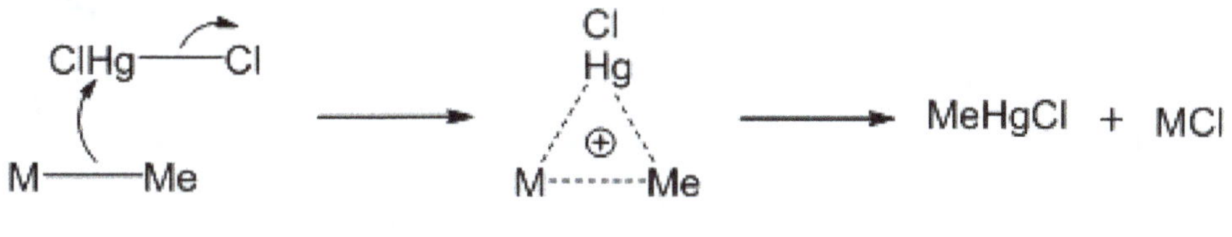

The retention of configuration proceeds by the following pathway.

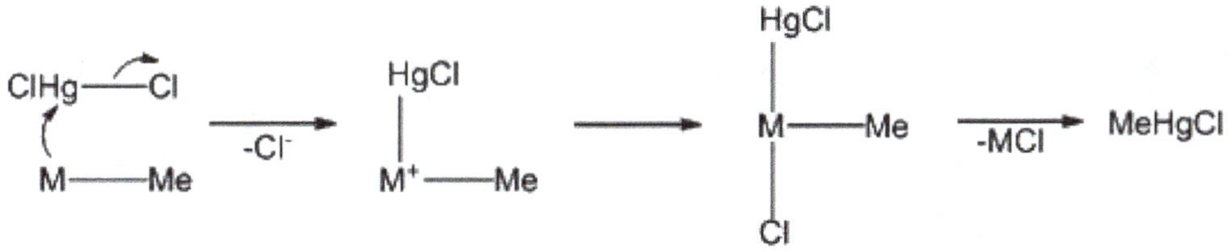

References

- Hartwig, J. F. (2010). Organotransition Metal Chemistry, from Bonding to Catalysis. New York: University Science Books. ISBN 1-891389-53-X

- Hall, Thomas L.; Lappert, Michael F.; Lednor, Peter W. (1980). "Mechanistic studies of some oxidative-addition reactions: free-radical pathways in the Pt^o-RX, Pt^o-PhBr, and Pt^{II}-R'SO$_2$X Reactions (R = alkyl, R' = aryl, X = halide) and in the related rhodium(I) or iridium(I) Systems". J. Chem. Soc., Dalton Trans. (8): 1448–1456. doi:10.1039/DT9800001448

- Johnson, Curtis; Eisenberg, Richard (1985). "Stereoselective Oxidative Addition of Hydrogen to Iridium(I) Complexes. Kinetic Control Based on Ligand Electronic Effects". Journal of the American Chemical Society. 107 (11): 3148–3160. doi:10.1021/ja00297a021

- Crabtree, Robert (2005). The Organometallic Chemistry of the Transition Metals. Wiley-Interscience. pp. 159–180. ISBN 0-471-66256-9

- Jay A. Labinger "Tutorial on Oxidative Addition" Organometallics, 2015, volume 34, pp 4784–4795. doi:10.1021/acs.organomet.5b00565

8

Organometallic Chemistry in Homogeneous Catalysis

Catalysis is the increase in the rate of chemical reaction caused due to the introduction of a substance called catalyst. The substance, however, does not directly participate in the process, but helps speed up the reaction. Catalysis can be classified as homogenous and heterogeneous. The former plays a vital role in application of organometallic chemistry. It is the form of catalysis in which the catalyst is of the same phase as the reactants. The topics elaborated in this chapter will help in gaining a better perspective about catalysis in organometallic chemistry.

Homogeneous Catalysis

In chemistry, homogeneous catalysis is catalysis in a solution by a soluble catalyst. Strictly speaking, homogeneous catalysis are catalytic reactions where the catalyst is in the same phase as the reactants, so homogeneous catalysis applies to reactions in the gas phase and even in a solid. Heterogeneous catalysis is the alternative to homogeneous catalysis, where the catalysis occurs at the interface of two phases, typically gas-solid. The term is used almost exclusively to describe solutions and it is often implies catalysis by organometallic compounds. The area is one of intense research and many practical apprehended applications, e.g., the production of acetic acid. Enzymes are examples of homogeneous catalysts.

Examples

Acid Catalysis

The proton is the most pervasive homogeneous catalyst because water is the most common solvent. Water forms protons by the process of self-ionization of water. In an illustrative case, acids accelerate (catalyze) the hydrolysis of esters:

$$CH_3CO_2CH_3 + H_2O \rightleftharpoons CH_3CO_2H + CH_3OH$$

In the absence of acids, aqueous solutions of most esters do not hydrolyze at practical rates.

Organometallic Chemistry

Processes that utilize soluble organometallic compounds as catalysts fall under the category of homogeneous catalysis, as opposed to processes that use bulk metal or metal on a solid support, which are examples of heterogeneous catalysis. Some well-known examples of homogeneous catalysis include hydroformylation and transfer hydrogenation, as well as certain kinds of Ziegler-Natta polymerization and hydrogenation. Homogeneous catalysts has also been used in a variety of industrial processes such as the Wacker process Acetaldehyde (conversion of ethylene to acetaldehyde) as well as the Monsanto process and the Cativa process for the conversion of MeOH and CO to acetic acid.

Many non-organometallic complexes are also widely used in catalysis, e.g. for the production of terephthalic acid from xylene.

Other forms of Homogeneous Catalysis

Enzymes are homogeneous catalysts that are essential for life but are also harnessed for industrial processes. A well studied example carbonic anhydrase, which catalyzes the release of CO_2 into the lungs from the blood stream.

Contrast with Heterogeneous Catalysis

Homogeneous catalysis differs from heterogeneous catalysis in that the catalyst is in a different phase than the reactants. One example of heterogeneous catalysis is the petrochemical alkylation process, where the liquid reactants are immiscible with a solution containing the catalyst. Heterogeneous catalysis offers the advantage that products are readily separated from the catalyst, and heterogeneous catalysts are often more stable and degrade much slower than homogeneous catalysts. However, heterogeneous catalysts are difficult to study, so their reaction mechanisms are often unknown.

Enzymes possess properties of both homogeneous and heterogeneous catalysts. As such, they are usually regarded as a third, separate category of catalyst.

Homogeneous Catalysis-I

One of the most important exploits of the organometallic chemistry is its application in the area of homogeneous catalysis. The field has now expanded its territory to accommodate in equal measures many large-scale industrial processes as well as numerous small scale reactions of the day-to-day organic synthesis. A few representative examples of organometallic catalysis are outlined below.

Alkene Isomerization

Alkene isomerization is a transformation that involve a shift of a double bond to an adjacent position followed by 1,3–migration of a H atom. The isomerization reaction is transition metal catalyzed.

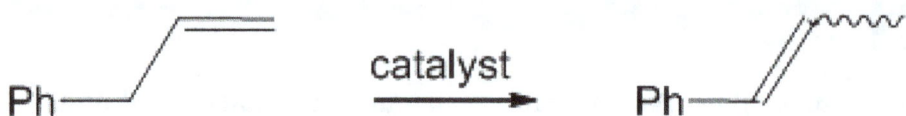

The alkene isomerization reaction may proceed by two pathways, (*i*) one through a η^1–alkyl intermediate and (*ii*) the other through η^3–allyl intermediate. In the η^1–alkyl pathway, an alkene first binds to a metal at a vacant site next to M–H bond and then subsequently undergoing an insertion into the M–H bond thus creating back the vacant site. The resultant species then undergoes a H atom transfer from the alkyl moiety to give the isomerized olefin along with the regeneration of the M–H species.

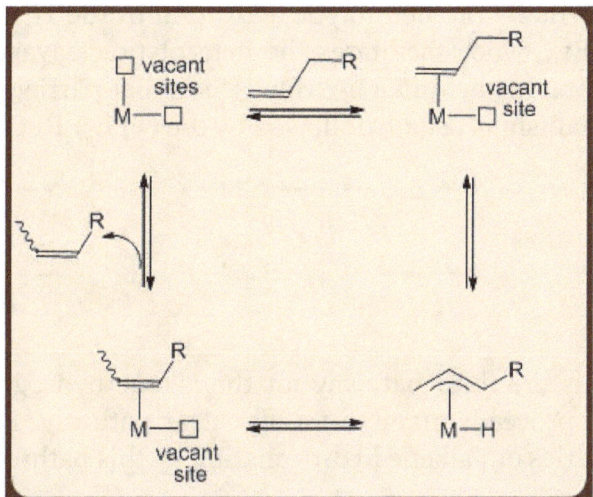

The η^3–allyl mechanism requires the presence of two vacant sites. This mechanism goes through a η^3–allyl intermediate formed by a C–H activation at the allylic position of the olefin formed after binding to the metal and alongside leads to the formation of a M–H bond. Subsequent H transfer from the metal back to the η^3–allyl moiety leads to the alkene isomerized product.

Alkene Hydrogenation

The transition metal catalyzed alkene hydrogenation reactions are of significant industrial and academic interest. These reactions involve the H_2 addition on a C=C bond of olefins to give alkenes. The alkene hydrogenation may proceed by three different pathways namely the (*i*) oxidative addition (*ii*) heterolytic activation and (*iii*) the homolytic activation of the H_2 molecule.

The oxidative addition pathway is commonly observed for the Wilkinson's catalyst $(PPh_3)_3RhCl$ and is the most studied among all of the three pathways that exist. The catalytic cycle initiates with the oxidative addition of H_2 followed by alkene coordination. The resultant species subsequently get converted to the hydrogenated product.

Wilkinson's catalyst

The second pathway proceeds by the heterolytic activation of the H_2 molecule and requires the presence of a base like NEt_3, which facilitates the heterolytic cleavage by abstracting a proton from the H_2 molecule and leaving behind a hydride H^- ion that participates in the hydrogenation reaction. This type of mechanism is usually followed by the $(PPh_3)_3RuCl_2$ type of complexes.

$$X—Y + M—Z \longrightarrow M—X + Y—Z$$

Homolytic cleavage of H_2 is the third pathway for the alkene hydrogenation. It is the rarest of all the three methods and proceeds mainly in a binuclear pathway. Paramagnetic cobalt based $Co(CN)_5^{3-}$ type catalysts carries out alkene hydrogenation by this pathway via the formation of the $HCo(CN)_5^{3-}$ species.

Arene Hydrogenations

Examples of homogeneous catalysts for arene hydrogenation are rare though it is routinely achieved using catalysts like Rh/C under the heterogeneous conditions. A representative example of a homogeneous catalyst of this class is $(\eta^3-allyl)Co[P(OMe)_3]_3$ that carry out the deuteration of benzene to give the all-*cis*-$C_6H_6D_6$ compound.

Transfer Hydrogenation

This is a new kind of a hydrogenation reaction in which the source of the hydrogen is not the H_2 molecule but an easily oxidizable substrate like isopropyl alcohol. The method is particularly useful for the reduction of ketones and imines but not very effective for the olefins.

$$Me_2CHOH \ + \ RCH{=}CH_2 \ \longrightarrow \ Me_2C{=}O \ + \ RCH_2CH_3$$

Homogeneous Catalysis-II

It is truly an exciting time for the field of organometallic chemistry as its potentials in homogeneous catalysis are being realized in an unprecedented manner. The growth in the field organometallic chemistry has been rightly acknowledged by the award of three Nobel prizes in over a decade in the areas of asymmetric hydrogenation (Nyori and Knowles in 2001), olefin metathesis (Grubbs, Schrock and Chauvin in 2006) and palladium mediated C–C cross coupling reactions (Suzuki, Negishi and Heck, 2010). A few representative examples of such landmark discoveries of homogeneous catalysis by organometallic compounds are discussed below.

Hydroformylation Reaction

Hydroformylation, popularly known as the "oxo" process, is a Co or Rh catalyzed reaction of olefins with CO and H_2 to produce the value-added aldehydes.

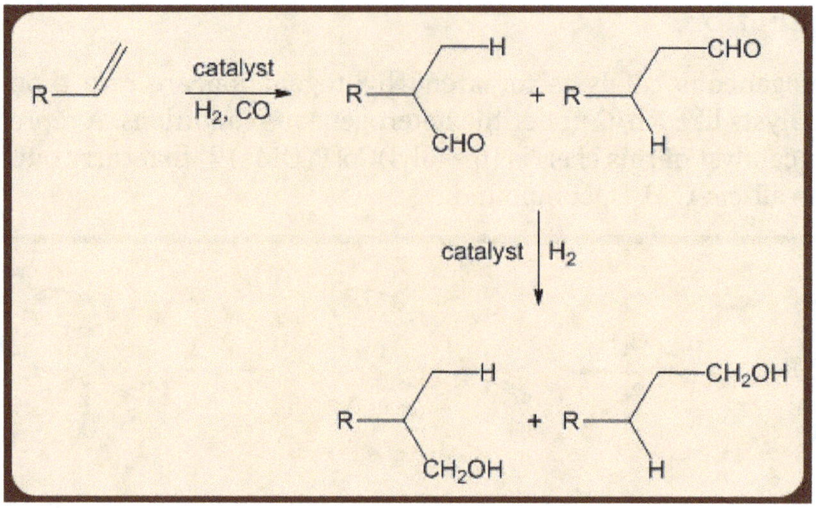

The reaction, discovered by Otto Roelen in 1938, soon assumed an enormous proportion both in terms of the scope and scale of its application in the global production of aldehydes. The metal hydride complexes namely, the rhodium based $HRh(CO)(PPh_3)_3$ and the cobalt based $HCo(CO)_4$ complexes, catalyzed the hydroformylation reaction as shown below.

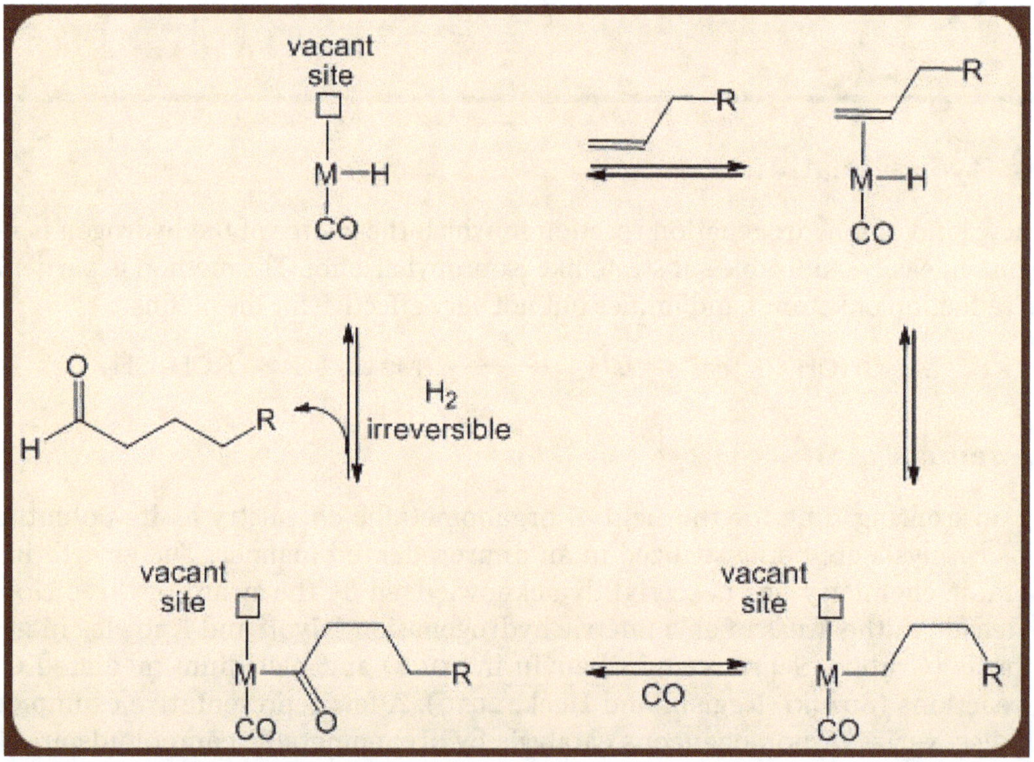

Transition Metal Oxo Complex

In coordination chemistry, an oxo ligand is an oxygen atom bound only to one or more metal centers. These ligands can exist as terminal or (most commonly) as bridging atom. Oxo ligands stabilize high oxidation states of a metal.

Oxo ligands are pervasive, comprising the great majority of the Earth's crust. This lesson concerns a subset of oxides, molecular derivatives. They are also found in several metalloenzymes, e.g. in the molybdenum cofactor and in many iron-containing enzymes. One of the earliest synthetic compounds to incorporate an oxo ligand is sodium ferrate (Na_2FeO_4) circa 1702.

Reactivity

Olation and Acid-base Reactions

Common reactions affected by metal-oxo compounds is olation, the condensation process that converts low molecular weight oxides to polymeric materials, including minerals. Olation often begins with the deprotonation of a metal-hydroxo complex.

Oxygen-atom Transfer

Oxygen-atom transfer is common reaction of particular interest in organic chemistry and biochemistry. Some metal-oxos are capable of transferring their oxo ligand to organic substrates. One such example of this type of reactivity is from and enzyme super-family Molybdenum oxotransferase.

Hydrogen Atom Abstraction

Transition metal-oxo's are also capable of abstracting strong C–H, N–H, and O–H bonds. Cytochrome P450 contains a high-valent iron-oxo which is capable of abstracting hydrogen atoms from strong C–H bonds.

Molecular Oxides

Some of the longest known and most widely used oxo compounds are oxidizing agents such as potassium permanganate ($KMnO_4$) and osmium tetroxide (OsO_4). Compounds such as these are widely used for converting alkenes to vicinaldiols and alcohols to ketones or carboxylic acids. More selective or gentler oxidizing reagents include pyridinium chlorochromate (PCC) and pyridinium dichromate (PDC). Metal oxo species are capable of catalytic, including asymmetric oxidations of various types. Some metal-oxo complexes promote C-H bond activation, converting hydrocarbons to alcohols.

Selection of molecular metal oxides. From left, vanadyl chloride (d^0), a tungsten oxo carbonyl (d^2), permanganate (d^0), [ReO_2(pyridine)$_4$]$^+$ (d^2), simplified view of compound I (a state of cytochrome P450, d^4), and trismesityliridium oxide (d^4).

Metalloenzymes

Iron(IV)-oxo Species

Oxygen rebound mechanism utilized by cytochrome P450 for conversion of hydrocarbons to alcohols via the action of "compound I", an iron(IV) oxide bound to a radical heme.

Iron(IV)-oxo compounds are intermediates in many oxidations catalysed by heme-containing enzymes. One of the most widely studied examples is cytochrome p450 enzymes, which use a hemecofactor that is capable of hydroxylation of saturated C–H bonds, epoxidation of olefins, and oxidation of aromatic groups. Similarly, methane monooxygenase (MMO) oxidizes methane to methanol via oxygen atom transfer from an iron-oxo intermediate at its non-heme di-iron center. First, C-H bonds are quite resistant to oxidation and are generally unreactive at moderate temperatures. Second, harsh oxidizing agents will generally oxidize an alcohol to a carboxylic acid, but these enzymes are able to oxidize an alkyl group to an alcohol without further oxidation to a carbonyl or carboxylic acid. The oxidant used in these enzymatic reactions is molecular oxygen in contrast with the harsh, toxic chemicals often found in conventional synthetic organic oxidations. As is generally the case with enzymatic reactions, these oxidations are chemically selective and take place at fast rates in aqueous solvent. Much of the effort in producing synthetic C-H bond activation catalysts has been inspired by these well designed natural catalysts.

Molybdenum/Tungsten Oxo Species

Three structural families of molybdenum cofactors: a) xanthine oxidase, b) sulfite oxidase, and c) (DMSO) reductase. The DMSO reductase features two molybdopterin ligands attached to molybdenum. They are omitted from the figure for simplicity. The rest of the heterocycle is similar to what is shown for the other two cofactors.

The oxo ligand (or analogous sulfido ligand) is nearly ubiquitous in molybdenum and tungsten chemistry, appearing in the ores containing these elements, throughout their synthetic chemistry, and also in their biological role (aside from nitrogenase). The biologically transported species and starting point for biosynthesis is generally accepted to be oxometallates MoO_4^{-2} or WO_4^{-2}. All Mo/W enzymes, again except nitrogenase, are bound to one or more molybdopterin prosthetic group. The Mo/W centers generally cycle between hexavalent (M(IV)) and tetravalent (M(VI)) states. Although there is some variation among these enzymes, members from all three families involve oxygen atom transfer between the Mo/W center and the substrate. Representative reactions from each of the three structural classes are:

- Sulfite oxidase: $SO_3^{-2} + H_2O \rightarrow SO_4^{-2} + 2\ H^+ + 2\ e^-$

- DMSO reductase: $H_3CS(O)CH_3$ (DMSO) $+ 2\ H^+ + 2\ e^- \rightarrow H_3CSCH_3$ (DMS) $+ H_2O$

- Aldehyde ferredoxin oxidoreductase: $RCHO + H_2O \rightarrow RCO_2H + 2\ H^+ + 2\ e^-$

Oxygen-evolving Complex

The active site for the oxygen-evolving complex (OEC) of photosystem II (PSII) is a Mn_4O_5Ca centre with several bridging oxo ligands that participate in the oxidation of water to molecular oxygen. The OEC is proposed to utilize a terminal oxo intermediate as a part of the water oxidation reaction. This complex is responsible for the production of nearly all of earth's molecular oxygen. This key link in the oxygen cycle is necessary for much of the biodiversity present on earth.

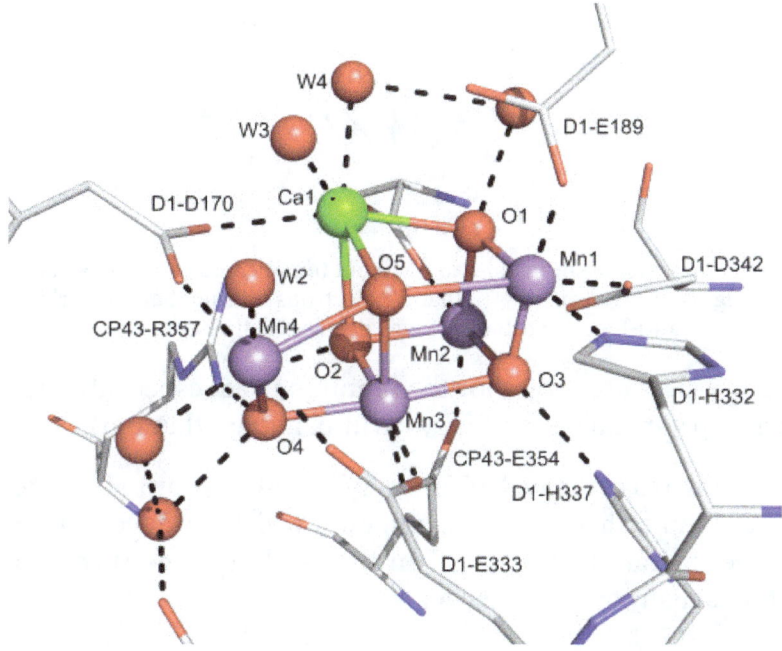

X-ray Crystal structure of the Mn_4O_5Ca core of the oxygen evolving complex of Photosystem II at a resolution of 1.9 Å.

The "Oxo Wall"

The term "oxo wall" is a theory used to describe the fact that no terminal oxo complexes are known for metal centers with tetragonal symmetry and d-electron counts beyond 5. Oxo compounds for the vanadium through iron triads (groups 3-8) are well known, whereas terminal oxo compounds for metals in the cobalt through zinc triads (groups 9-12) are rare and invariably feature metals with coordination numbers lower than 6. This trend holds for other metal-ligand multiple bonds. Claimed exceptions to this rule have been retracted.

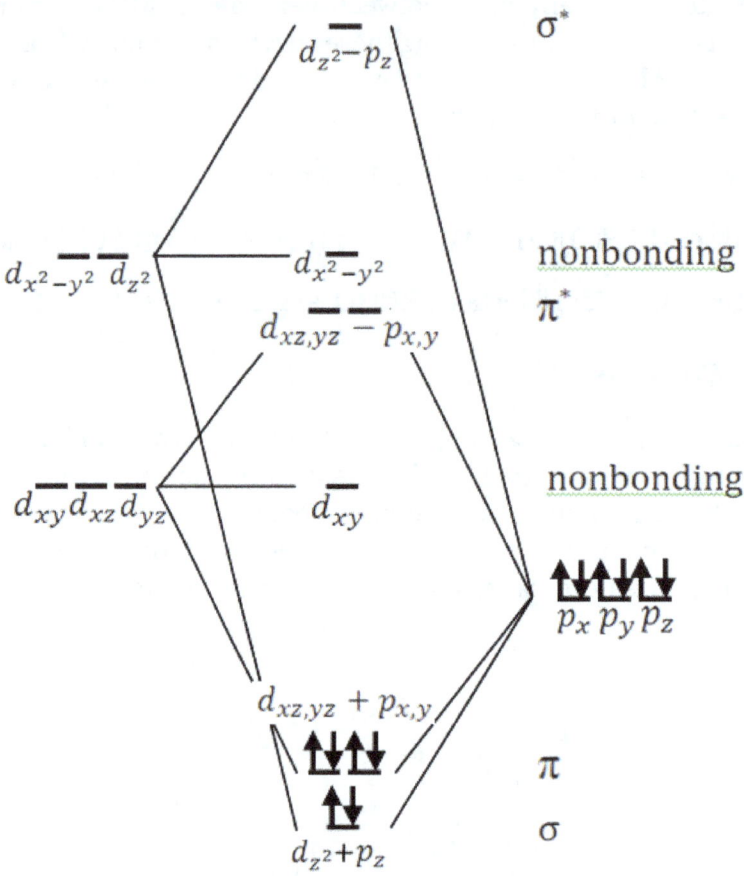

Qualitative molecular orbital diagram of a d^0 metal-oxo fragment (empty metal d orbitals in an octahedral field on left, full oxygen p orbitals on right). Here it can be seen that d^{1-2} electrons fill a nonbonding orbital and electrons d^{3-6} fill anti-bonding orbitals, which destabilize the complex.

Terminal oxo ligands are also rather rare for the titanium triad, especially zirconium and hafnium and is unknown for group 3 metals (scandium, yttrium, and lanthanum).

At first glance, the iridium oxo complex $Ir(O)(mesityl)_3$ may appear to be an exception to the oxo-wall, but it is not. The complex has trigonal symmetry, which produces a reordering of the metal d-orbitals below the degenerate MO pi* pair. In three-fold symmetric complexes, multiple MO bonding is allowed for as many as 7 d-electrons.

Palladium-catalyzed Coupling Reactions

Palladium-catalyzed coupling reactions comprise a family of cross-coupling reactions that employ palladium complexes as catalysts. It is an active area of research and applications in homogeneous catalyst. In 2010, the Nobel Prize in Chemistry was awarded to Richard F. Heck, Ei-ichi Negishi and Akira Suzuki for their work on palladium-catalyzed cross couplings in organic synthesis.

Sonogashira coupling reaction mechanism

Examples

The reactions generally obey the following stoichiometry:

$$X\text{-}R + M\text{-}R' \rightarrow MX + R\text{-}R'$$

Variations are based on the identity of X-R (often an aryl bromide) and M-R'. Often the reactions generate salts, or salt-like products (zinc halides, tin halides, silicon halides)

- Negishi coupling between an organohalide and an organozinc compound

- Heck reaction between alkenes and aryl halides

- Suzuki reaction between aryl halides and boronic acids

- Stille reaction between organohalides and organotin compounds

- Hiyama coupling between organohalides and organosilicon compounds

- Sonogashira coupling between aryl halides and alkynes, with copper(I) iodide as a co-catalyst

- The Buchwald-Hartwig amination of an aryl halide with an amine, extended to aryl halide with phenol and thiol

- The Kumada coupling of grignards and aryl or vinyl halides

- The Heck-Matsuda Reaction of an arenediazonium salt with an alkene

Catalysts

Typical palladium catalysts used include the following compounds:

- palladium acetate, $Pd(OAc)_2$

- tetrakis(triphenylphosphine)palladium(0), $Pd(PPh_3)_4$

- bis(triphenylphosphine)palladium(II) dichloride, $PdCl_2(PPh_3)_2$

- [1,1'-bis(diphenylphosphino)ferrocene]palladium(II) dichloride

Some of these catalysts are really pro-catalysts, that become activated in situ. For example, Pd-$Cl_2(PPh_3)_2$ is reduced to a Pd(0) complex or transmetalated to a Pd(II) aryl complex before it participates in the catalytic cycle.

Operating Conditions

Unoptimized reactions typically use 10-15 mol% of palladium. In optimized reactions, catalyst loadings can be on the order of 0.1 mol % or below. Palladium nano clusters have been found to catalyze coupling reactions with catalyst loadings as low as parts per billion, however such systems typically do not maintain catalytic activity as long as well defined ligated catalysts. Many exotic ligands and chiral catalysts have been reported, but they are largely not available commercially, and do not find widespread use. Much work is being done on replacing the phosphine ligands with other classes, such as Arduengo-type carbene complexes, as the phosphine ligands are typically oxygen sensitive (easily oxidized) and must be handled under an inert atmosphere. Phosphines are labile, sometimes requiring additional ligand. For example, $Pd(PPh_3)_4$ would be supplemented with PPh_3 to keep the palladium coordinated despite loss of the labile phosphine ligands.

A concern with the use of palladium in the preparation of pharmaceuticals is that traces of the toxic heavy metal will remain in the product. Column chromatography can be used, but solid-phase metal scavengers (ion exchange resins and derivatives of silica gel) promise more efficient separation.

C–C Cross-coupling Reactions

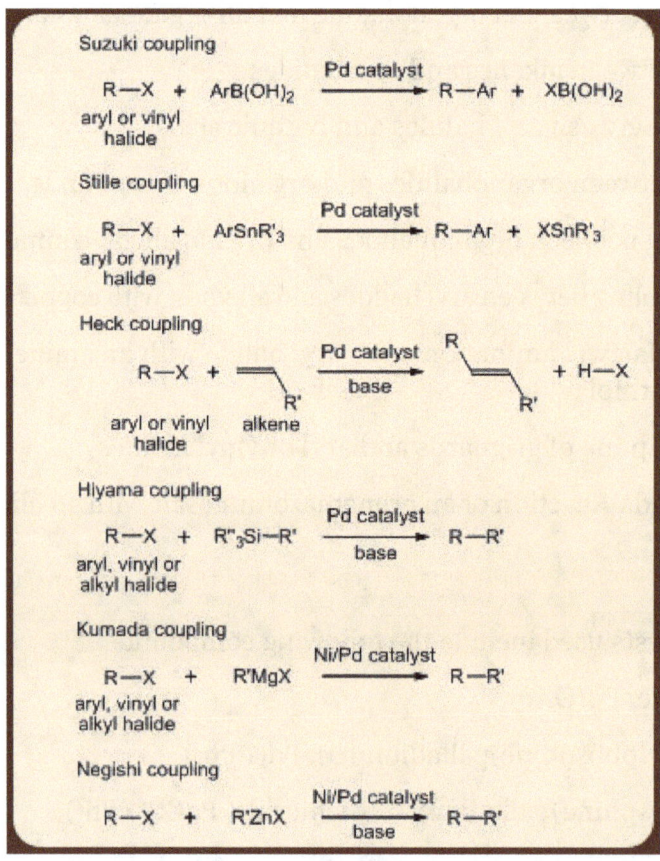

Various types of the palladium mediated C–C cross-coupling reactions.

The palladium catalyzed cross-coupling reactions are a class of highly successful reactions with applications in the organic synthesis to have emerged recently. The reactions carry out a coupling of the aryl, vinyl or alkyl halide substrates with different organometallic nucleophiles and as such encompasses a family of C–C cross-coupling reactions that are dependent on the nature of nucleophiles like that of the B based ones in the Suzuki-Miyuara coupling, the Sn based ones in the Stille coupling, the Si based ones in the Hiyama coupling, the Zn based ones in the Negishi coupling and the Mg based ones in the Kumada coupling reactions.

An unique feature of these reactions is the exclusive formation of the cross-coupled product without the accompaniment of any homo-coupled product. Another interesting feature of these coupling reactions is that they proceed via a common mechanism involving three steps that include the oxidative addition, the transmetallation and the reductive elimination reactions.

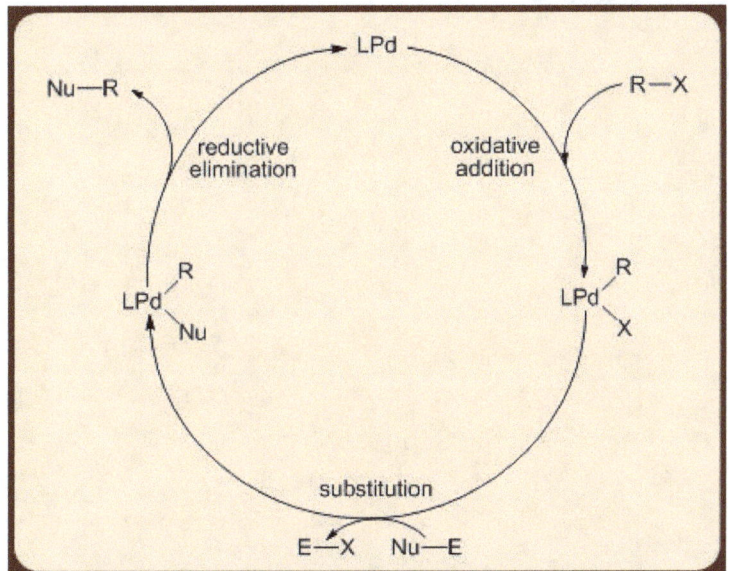

A general catalytic cycle for the palladium mediated C–C cross-coupling reactions.

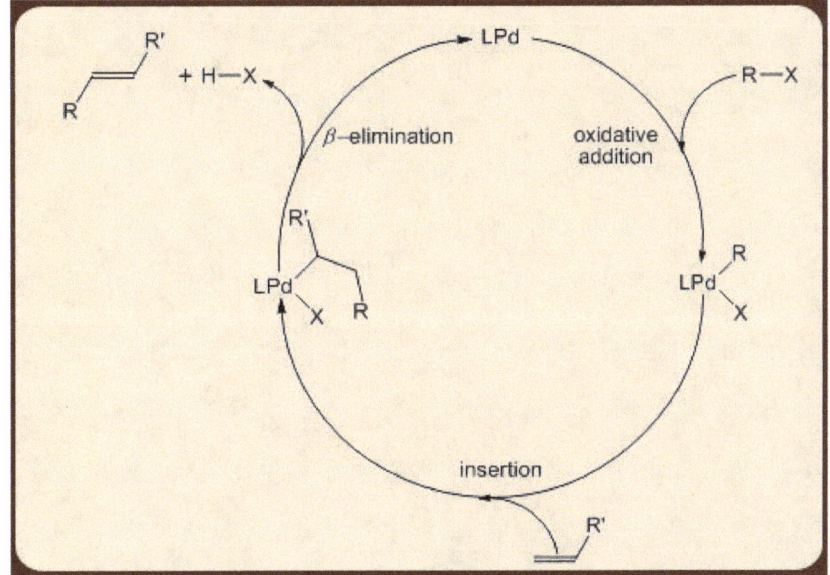

A catalytic cycle for the palladium mediated Heck coupling reaction.

References

- Schwarz, G., Mendel, R.R., and Ribbe, M.W. (2009). "Molybdenum cofactors, enzymes and pathways". Nature. 460 (7257): 839–847. Bibcode:2009Natur.460..839S. doi:10.1038/nature08302

- Korzekwa, Kenneth; Trager, William; Gouterman, Martin; Spangler, Dale; Loew, Gilda (1985). "Cytochrome P450 mediated aromatic oxidation: a theoretical study". Journal of the American Chemical Society. 107 (14): 4273–4279. doi:10.1021/ja00300a033. ISSN 0002-7863

- Winkler, J. R.; Gray, H. B. (2012). "Electronic Structures of Oxo-Metal Ions". Struct. Bond. Structure and Bonding. 142: 17–28. doi:10.1007/430_2011_55. ISBN 978-3-642-27369-8

- Holm, R. H. (1987). "Metal-centered oxygen atom transfer reactions". Chem. Rev. 87 (6): 1401–1449. doi:10.1021/cr00082a005

- Gunay A. & Theopold, K.H. (2010). "C-H Bond Activations by Metal Oxo Compounds". Chem. Rev. 110 (2): 1060–1081. doi:10.1021/cr900269x

Permissions

Index

www.ingramcontent.com/pod-product-compliance
Lightning Source LLC
Chambersburg PA
CBHW080408190526
45161CB00003B/169